Development and Characterization of AA7075 Aluminum Surface Nanocomposites

By
Nirma

TABLE OF CONTENTS

CHAPTER 1

INTRODUCTION

1.1 BACKGROUND OF RESEARCH

AA7075-T6 is a high strength aluminium alloy with an excellent strength-to-weight ratio, making it highly preferable for aerospace, automotive, marine, defense, and structural applications (Altas et al. 2024). The merits are low density, high specific strength, good machinability, and suitability for load bearing components. The poor tribological and corrosion performance of AA7075 alloy limits its use in severe service conditions. The drawbacks can be addressed through surface modification techniques. Friction Stir Processing (FSP) is an advanced surface modification technique that modifies the microstructure of materials, improving mechanical properties, wear resistance, and surface integrity (Butola et al. 2022a). Unlike shot peening, anodizing or laser surface treatment, which affect only the surface layer, FSP refines the entire processed zone. It improves surface properties on aluminium alloys in a better way, unlike anodizing. By mechanically mixing the material, FSP refines grain structure, enhances strength, incurs low cost and is environmentally friendly, requiring no chemicals or external energy sources.

Stir casting is another cost-effective route for developing nanocomposites but often struggles with nanoparticle agglomeration at high concentrations (Mishra et al. 2025). In contrast, FSP provides superior control over particle distribution, leading to ultra-fine-grained microstructure, enhanced mechanical and surface properties of alloys like AA7075. FSP is widely employed to fabricate surface nanocomposites by adding appropriate reinforcements into the base matrix material through grooves or blind holes. The addition of reinforcements during FSP further enhances surface performance (Raja et al. 2020). Carbonaceous materials include graphene (Gr), graphene oxide (GO) and graphene nanoplatelets (GNPs), along with ceramic reinforcements such as silicon carbide (SiC) are most appropriate candidate materials for surface modification owing to their unique properties. Gr exhibited excellent tribological and self-lubricating behaviour. GO, with its layered structure and functional groups, reduces corrosion rate and enhances strength by improving the bonding and passivation of the surface. GNPs are ideal for applications demanding high thermal conductivity, exceptional strength, enhanced wear resistance and mechanical stability (Wang et al. 2024).

SiC is ideal for applications requiring superior hardness, high thermal stability, excellent wear resistance and enhanced load-bearing capability. In comparison, traditional reinforcements such as titanium diboride (TiB_2) and boron nitride (BN) do not offer the same level of performance for high-performance applications involving AA7075-T6 (Sowrabh et al. 2021). While TiB_2 exhibits some hardness and wear resistance, it lacks the flexibility and thermal conductivity of ceramic and carbon-based reinforcements. BN, though useful for lubrication, does not significantly contribute to mechanical or surface properties in high temperature conditions. Another issue is that achieving uniform dispersion and good bonding of BN with the aluminium matrix can be challenging, further limiting its effectiveness.

The surface characteristics of friction stir processed (FSPed) surface depends on reinforcement nature, the amount it contains (volumetric or weight %), morphology, distribution with matrix material, the size of agglomerates and the quality of interface with matrix material. To achieve desired surface property, FSP variables must be carefully selected (Ghadar et al. 2021). The Tool Rotational Speed (TRS) and Tool Traverse Speed (TTS) are significantly affected heat input, strain rate and temperature, those are essential for dynamic recrystallization (DRX). In addition, tool profile (TP) influenced the reinforcement distribution within the matrix material, frictional heat generation, defect formation, surface finish and material mixing. Incorporation of carbonaceous and ceramic reinforcements on the surface of AA7075-T6 by employing FSP with optimum selection of TTS, TRS and TP can be promise in the enhancement of metallurgical, mechanical, surface topography, corrosion resistance and tribology property.

1.2 MOTIVATION OF RESEARCH

The maximum temperature of AA7075-T6 alloy can withstand during extreme flight speed depends on aerodynamic heating and thermal protection measures. AA7075-T6 maintains mechanical integrity upto 120^0C for prolonged durations. At temperature between 200^0 and 250^0 C, strength loss takes place results in gradual decrease in tensile properties. For instance, the temperature ranges from 25^0C to 400^0C, alloy sensitivity to thermal exposure takes place. When operating -100^0C to 300^0C, coalescence of particles and precipitates took place. At extreme speeds (supersonic or hypersonic) aerodynamic heating can cause surface temperature to exceed 250^0C, potentially affecting alloy performance (Smith 2021). The Concorde, a Mach 2.04 supersonic passenger jet, experienced 127^0C on its skin at cruising speed and it requires specially designed aluminium-copper-magnesium alloy (RR58) to withstand this heat while maintaining structural integrity. Since AA7075-T6 performs well in subsonic and transonic flight conditions, on other hand, supersonic or hypersonic conditions require alternative materials. Also, the parts used in turbines, propulsion systems used in rockets, missiles and aircraft, fasteners used in aero-engine systems, and tooling systems for the hot working process should have experienced material degradation problems around the temperature ranges between 400 and 700^{0}C (Jia and Gu 2014; McLouth et al. 2018; Wang et al. 2012). From the applications above, the elevated temperature withstanding capabilities in mechanical, corrosion and tribological performances are much needed.

To improve elevated temperature performance of AA7075-T6, the surface can be modifying with ceramic (SiC) and carbonaceous (Gr, GNPs and GO) reinforcements. The addition of such reinforcements to aluminium alloys has been the subject of numerous investigations. In addition to the reinforcement type (RT), FSP machine variable (TTS and TRS) and tool profile (TP) play a crucial role in determining the extent of plastic deformation, thermal exposure, surface finish, defect formation and reinforcement dispersion within the stir zone (SZ). However, a systematic work exploring the impact of unique reinforcements in conjunction with TRS, TTS and TP on the comprehensive performance of AA7075-T6 is lacking. Furthermore, it has not been thoroughly investigated how carbon-based reinforcements and ceramic reinforcements differ in performance in the AA7075-T6 alloy. This research aims to fill this gap by evaluating the impact of TTS, TP and reinforcements on properties of AA7075-T6, such as microstructure, mechanical, corrosion resistance and tribological. Based on outcomes, the optimal combinations of aforementioned variables for high performance applications are recommended. Fig.1.1 shows the proposed research scheme.

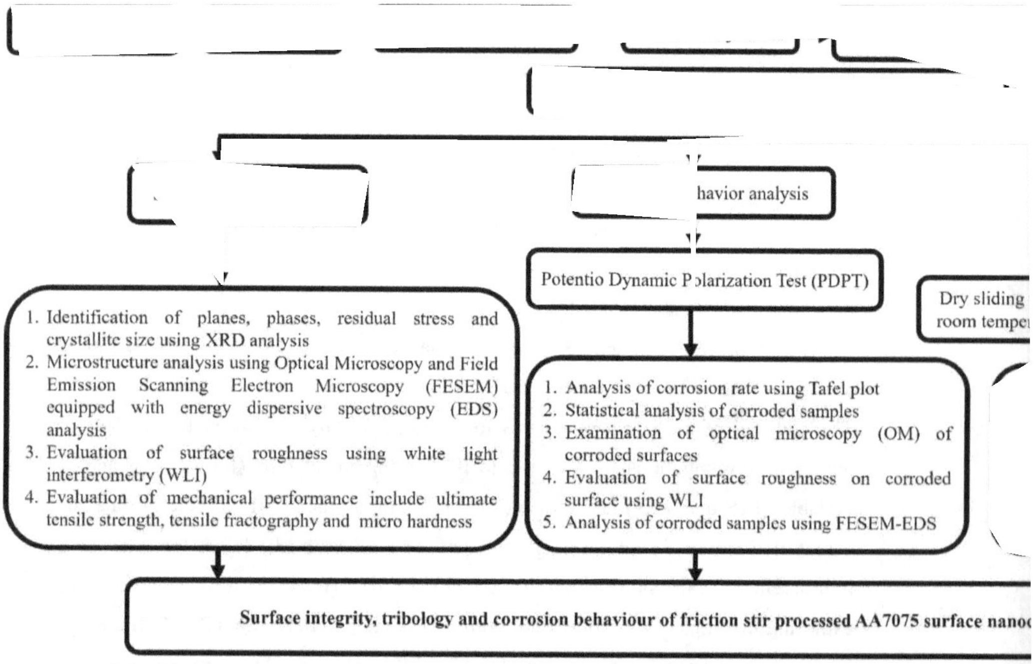

Fig.1 Proposed research scheme for friction stir processing on AA707

1.3 OBJECTIVES AND SCOPE OF RESEARCH

- Modification of AA7075-T6 surface using friction stir processing with unique combination of reinforcements, tool traverse speed and tool profile.
- Conduct a thorough assessment of the of tool traverse speed on temperature distribution and residual stress.
- Examine the effect of tool traverse speed on surface integrity of AA7075 surface nanocomposites in terms of surface roughness, crystallite size, lattice strain, ultimate tensile strength and microhardness.
- Evaluate the impact of tool traverse speed, tool profile and reinforcement on corrosion behaviour of aluminium surface nanocomposites
- Conduct a thorough assessment of the impact of tool traverse speed, tool profile and reinforcement on tribological behaviour of AA7075 surface nanocomposites in terms of specific wear rate, coefficient of friction and wear mechanism.

1.4 THESIS ORGANIZATION

Chapter 2 details a literature survey on AA7075-T6, including its applications, associated problems, and surface modification routes. It also detailed friction stir welding (FSW), need for choosing friction stir processing (FSP), influence of FSP variables on temperature distribution, metallurgical, mechanical, surface integrity, corrosion and tribological behaviour. Chapter 3 details the fabrication and characterization techniques involved in the development of aluminium surface nanocomposites. Chapter 4 investigates the influence of tool traverse speed (TTS) on surface integrity of aluminium surface nanocomposites in terms of microstructure, crystallite size, lattice strain, surface roughness, UTS and microhardness. Chapter 5 deals with the effect of TTS and reinforcement type (RT) on electrochemical corrosion behaviour of aluminium surface nanocomposites. Chapter 6 investigates the influence of TTS and RT on tribological behaviour of aluminium surface nanocomposites. Chapter 7 details the conclusion and future scope of work.

CHAPTER 2
LITERATURE REVIEW

2.1 INTRODUCTION

AA7075-T6 is a precipitation-hardenable alloy, known for its highest strength and lowest weight, specifically used in aircraft fuselage. This alloy face poor yield strength, stress corrosion cracking (SCC), deformability, and tribological characteristics. Surface modification processes are being explored to overcome these difficulties. Friction stir processing (FSP) is a surface modification technique, emerging low-cost, environment-friendly and able to produce ultrafine-grained microstructure. This chapter reviews AA7075-T6 alloy merits, demerits, and applications requirements. It also deals why bulk modification is not prefer over surface modification and why FSP is chosen. This chapter further discusses the improvement in surface properties by incorporating carbonaceous and ceramic reinforcements to AA7075-T6. The details about sample preparation, FSP tool, and machine variables are included. The influence of machine variables on generation of heat during FSP is discussed in detail. The emphasis of surface integrity analysis is highlighted. Influence of grain size, tool traverse speed, and reinforcement type on corrosion rate are explained, and finally, the need for tribological test under different conditions is discussed.

2.2 AA7075-T6 ALLOY

AA7075-T6 is a high strength aluminium alloy with an excellent strength-to-weight ratio, making it highly preferable for aerospace, automotive, marine, defense, and structural applications. The first two digits (AA), third digit (7), fourth digit (0), last two digits (75) and T6, represents, Aluminium Association, series identifier, alloy modification, specific alloy and tempered condition, respectively. The main alloying element in this aluminium series is zinc (Zn). The digit "0" denotes original alloy since it shows no discernible change. The next two numbers, "75," denote the composition of a particular alloy (Altas et al. 2024), where zinc (Zn) ranges from 5.6 to 6.1%, magnesium (Mg) ranges from 2.1 to 2.9%, copper (Cu) ranges from 1.2 to 2%, and the balance is aluminium (Al). Cu contributes to enhanced hardness and strength. Precipitation sequence of this alloy as follows: Super saturated solid solution (SSS) → Guinier Preston (GP) zones → eta prime (η') → eta (η) ($MgZn_2$). In this sequence, the main precipitate phase is η', while the equilibrium phase is η ($MgZn_2$). The precipitation behavior significantly dependent on chemical composition of Cu. When the Cu content in the 7XXX series exceeds one weight percentage (wt.%), Cu atoms migrate into the

GP zones, exhibiting exceptional strength (De and Mishra 2011; Feng et al. 2013; Yadav and Bauri 2012a).

The alloying elements Cu, Mg and Zn play a crucial role in solution heat treatment (SHT), as they enhance the precipitation of intermetallic compounds. Another main advantage of this alloy is high strength to weight ratio. This ratio, refers to as specific strength (material strength relative to its weight), which indicates how much load a material can bear for its weight. This property is crucial where applications required light weight without sacrificing strength. The specific strength of AA7075-T6 is approximately four times greater than steel, which implies for the same weight, AA7075-T6 can support four times the load compares to steel. Using AA7075-T6 for fuselage of aircraft significantly reduces overall weight while maintaining the necessary strength to withstand high aerodynamic and structural loads. In contrast, a steel fuselage of the same strength would weigh nearly four times heavier, resulting in enhanced fuel consumption and declined payload capacity.

2.2.1 Application and requirements of aerospace aluminium series

Zinc based aluminium alloy widely used in aviation, aerospace, automobile, shipbuilding, consumer goods, defence, military vehicles, extrusion parts, electronic sector and stamp forging. Owing to its high strength-to-weight ratio, airframe structures constructions are typically preferred this Zn based aluminium alloys. It utilizes airframe structures include upper wing covers, spars, ribs, fuselage, attachments of wing to fuselage. The structural demands of modern jetliners and military transport aircraft align closely with the performance characteristics of Zn based aluminium alloys, making them alternative material choice in critical load-bearing components. Fig.2.1 displays the various sections of an aircraft require tailored properties such as fatigue resistance (FAT), fracture toughness (FT), corrosion resistance, compressive yield strength (CYS), and tensile strength (TS). The Zn based aluminium alloys are well-suited to meet these multi-property requirements across several major airframe components as follows (Staley and Lege 1993).

i. Wing structures: The upper and lower wing skins, spars, and stringers are subject to alternating loads during flight, necessitating high resistance to fatigue crack growth (FCG), TS, and Young's modulus (E). The Fig.2.1 shows the importance of these properties. Alloys such as AA7075-T6 and AA7475-T761 are frequently used in these components because of their proven resistance to fatigue and cracking.

ii. Fuselage: The fuselage skins, frames, and stringers experience complex loading and pressurization cycles, making fatigue resistance and corrosion protection crucial. The Fig.2.1 shows these properties prominently, FCG, TS, E, and corrosion resistance listed as design requirements.

iii. Wing to Fuselage Attachments: These joints must accommodate both shear and compressive loads. The Fig.2.1 denotes these areas as requiring high FT, FAT, and TS, which are key strengths of the Zn based aluminium series. Their high static strength ensures structural integrity under extreme flight conditions.

iv. Control Surfaces and Stabilizers: The horizontal stabilizers, particularly under tension and compression, are marked in the Fig.2.1 with requirements for FT, FCG, and TS.

v. Floor Beams and Cargo Tracks: These internal components demand TS and resistance to environmental degradation.

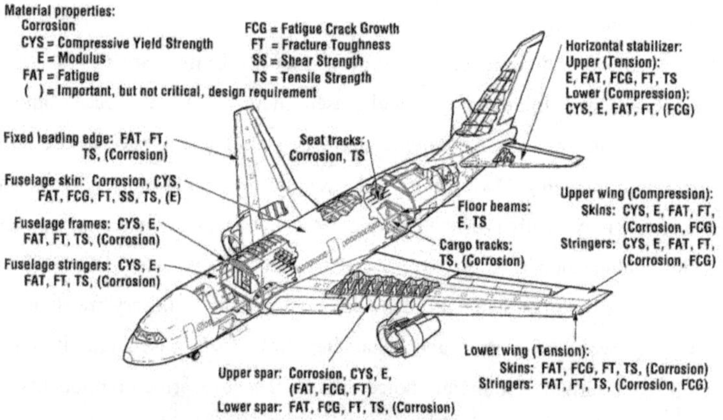

Fig.2.1 Property requirements for Jetliners and Military Transport aircraft applications (Staley and Lege 1993)

Overall, the mapping of specific mechanical property requirements to different airframe components, as depicted in the Fig. 2.1, demonstrates a strong correlation with the capabilities of Zn based aluminium alloys. Their sustained application in both military aviation and commercial reflects their proven reliability in meeting the required demands of aerospace structural design.

2.2.2 Constraints and Issues of AA7075-T6 alloy

While very much in recent usage and exhibiting satisfactory mechanical characteristics, AA7075-T6 alloy also has several demerits as follows

i. It exhibits severe SCC, when it is exposed to high humidity and marine environment. The constituent particles (CPs) of Cu induced such kind of corrosion.
ii. It also subjected to localized corrosion, such as galvanic, intergranular attack, pitting or dealloying, due to the existence high Zn content,
iii. The excessive amount of Mg leads to hot cracking.
iv. The excessive amount of Zn leads to reduction in ductility.
v. It exhibits low weldability, hot cracking susceptibility, and considerably lower mechanical properties in the heat-affected zone (HAZ), due to its chemical composition and heat-treatment condition.
vi. When exposed to harsh conditions and cyclic loading, the long-term characteristics of AA7075-T6 may deteriorate. It has been determined that the alloy would experience fatigue cracking, which will be most noticeable in services with significant vibration and weight.
vii. It has poor tribological qualities owing to its comparatively low hardness value, which shortens the service life.
viii. It has difficult to achieve a fine surface finish owing to its tendency toward tool wear and surface tearing when machined, further affects service life of the component.

Compared to other aluminium alloys, AA7075-T6 is costly due to its alloying elements and manufacturing process. Furthermore, a minimum ductility when in T6 condition also limits its shapeability, which exhibits in difficulty shaping complex geometries without cracking. Solution heat treatment (SHT) following forming process (casting), together with artificial aging requires most controlling parameter, makes up the T6 temper. Improper selection of heat treatment can lead to poor mechanical properties, resulting in loss of strength and toughness. Table 2.1 displays a comparison of tempering conditions such as T6, T7, and T8, highlighting their influence on SCC, mechanical strength, and corrosion resistance. The T6 condition involves SHT followed by aging at a specific temperature, resulting in high strength but enhanced susceptibility to SCC and poor corrosion resistance. In contrast, the T7 condition (Dixit et al. 2008), characterized by an overaged structure with evenly distributed precipitates, offers superior resistance to corrosion and reduced the risk of SCC, though it provides only moderate strength. The T8 condition, which exhibits a similar treatment to T6 but at a higher aging temperature, serves as a balance between the two; it exhibits strength slightly lower than T6 yet greater than T7, along with improved corrosion resistance and moderate resistance to SCC. This comparison is essential in choosing the appropriate tempering condition based on the desired balance between strength and durability in corrosive environments. Owing to the merits of the T8 condition, the authors recommended it as an alternative choice to the T6 and T7 conditions (Pankade et al. 2018; Sunar et al. 2020).

Table 2.1. Tempering condition versus corrosion and strength

Condition	Description	Influence on SCC	Nature of strength	Nature of corrosion resistance
T6	Solution heat treatment followed by aging at specific temperature	More	High	Poor
T7	Overaged and precipitates are evenly distributed	Reduced	Moderate	Superior
T8	Treatment is similar with T6 but higher aging temperature	Between T6 and T7	Slightly lower than T6, but higher than T7	Good

While conventional SHTs impose a trade-off between strength and corrosion resistance, surface modification processes provide a pathway to selectively increase surface properties (corrosion and tribology behavior) without compromising bulk properties (core remains tough).

2.2.3 Surface modification techniques

Surface modification of metallic materials has emerged as a crucial field of study, to enhance surface characteristics including fatigue life, hardness, corrosion resistance, and wear resistance without changing the bulk properties of the substrate (Ramezani et al. 2023). Numerous methods for integrating or depositing reinforcements into metals' near-surface areas have been devised. These consist of laser processing, plasma spraying, electron-beam surface melting, friction stir processing (FSP) and approaches based on powder metallurgy (PM). A high-energy laser beam is used in laser processing procedures (such as laser melt deposition and laser cladding) to create a melt pool where ceramic or metallic reinforcements can be added (Li et al. 2022; S. et al. 2023). This makes it possible to create metallurgically bonded coatings that are more resistant to corrosion and wear. The coating's functional qualities are endangered by the high localized temperatures, which frequently encourage undesired interfacial reactions like carbide formation or reinforcement dissolution. Additionally, laser setups are costly and necessitate exact process parameter control.

Thermal spray methods such as plasma spray, high-velocity oxy fuel spraying involved melting or semi-melting feedstock materials, which were propelled towards the substrate (Kumar et al. 2021). This technique allows deposition over wide areas and produced coatings with good adhesion. Nevertheless, porosity, poor particle-matrix bonding, and the requirement for substantial post-processing were the common problems with these coatings. Electron-beam surface melting is being used to incorporate reinforcements into a metallic surface. It forms hard intermetallic or ceramic phases in situ by melting a layer of powder that was previously applied to the substrate. Large scale industrial use is restricted by the need for costly electron accelerators and vacuum operation, despite the potential for enhanced hardness and wear resistance. PM or cast sintering techniques had been explored, in which reinforcement particles were mixed with metal powders or molten alloys before compaction/sintering or solidification. Although dense bulk composites can be developed using these techniques, agglomeration, segregation, or reactions at high temperatures make it challenging to control the distribution of nano-reinforcements. FSP has become a desirable substitute to address the aforesaid backdrops. It is a variant of Friction Stir Welding (FSW), incurs low-cost, environment-friendly and effectively refines surface microstructures and redistributes precipitates, thereby enhancing surface properties while retaining high mechanical properties (Nabi et al. 2025). Compared to melt-based surface modification techniques, FSP offers clear advantages owing to its solid-state nature. SiC reinforcements frequently decompose or react with the molten matrix in laser or plasma spray techniques, produced unwanted products like TiC or Ti_5Si_3 in Ti-based alloys that could embrittle the composite. FSP is an alternative route, reduces these reactions and maintains the inherent qualities of the reinforcing by operating below melting temperatures. Microstructural refinement is another notable feature.

2.3 FRICTION STIR PROCESSING

Friction stir processing (FSP), derived from friction stir welding (FSW), is a highly effective technique for surface modification with the production of ultrafine-grained microstructure (Lonavath and Boda 2023; Reyaz and Sinha 2024; Satyanarayana et al. 2024). Surface composites were developed by adding reinforcement particulates through pre-exiting groove/holes on the base metal (Maji et al. 2020). The groove is being cut on the base material with 160 mm, 2 mm and 1.5 mm of length, width and depth, respectively (Butola et al. 2022b). The details of surface preparation prior to FSP, as displayed in Table 2.2. The groove/hole was later processed using FSP tool which is made up of different materials (Butola et al. 2023; Dhanesh Babu et al. 2021; Elangovan and Balasubramanian 2007).

Table 2.2 Surface preparation prior to FSP for various matrix surface

Reference	Matrix surface	Groove details (mm)		Hole details (mm)	
		Width	Depth	Diameter	Depth
(Butola et al. 2020)	AA7075-T6	-	-	2	1.5
(Maji et al. 2021)	AA7075	2	2	-	-
(Jalilvand et al. 2019)	A356	-	-	2	4
(Khayyamin et al. 2013)	AZ91	1	2	-	-
(Kang et al. 2025)	WE43 Mg alloy	-	-	2	4
(Arulmoni et al. 2015)	Cu	1	2	-	-

The FSP tool contains shoulder diameter, pin diameter, pin length and pin geometry. The different pin geometries of FSP tool profiles were reported (Butola et al. 2022b; Elangovan and Balasubramanian 2007; Satheesh et al. 2020). Table 2.3 displays the details of FSP tool used in the previous studies. Among FSP tool materials, H13 tool holds good hardness and ability to withstand high mechanical stress and temperature during processing (Biradar et al. 2025). Researchers found that pin geometry has a major impact on material flow, with threaded pins producing better material mixing than smooth cylindrical pins (Akbari et al. 2025).

Table 2.3 Details of various FSP tools employed during processing

Reference	Tool material	Tool shoulder diameter (mm)	Tool pin diameter (mm)	Tool pin profile
(Kim et al. 2017)	SKD61	16	4	Screw
(Tekiyeh et al. 2019)	H13	20	6	Cylindrical and square
(Moustafa et al. 2020)	K110 Tool steel	25	8	Triangle
(Rana et al. 2016)	M2 grade tool steel	18	5.0-3.0	Cylindrical Taper
(Maji et al. 2021)g10	HCHCr Steel	18	6	Threaded

During FSP, friction generates heat between the tool shoulder and the workpiece surface. The heat generation is directly proportional to tool rotational speed (TRS) and inversely proportional to TTS (Elangovan and Balasubramanian 2007; Khodabakhshi et al. 2020; Netto et al. 2018). The flow stress is reduced at higher temperatures, and the material becomes soft. A tool pin inserted at the bottom of the shoulder stirs the softened material from the advancing side (AS) to the retreating side (RS). The material flows around the rotating action of the pin and is subjected to severe plastic deformation (SPD) and thermal exposure, which leads to significant refinement of the microstructure. The schematic of FSP is shown in Fig.2.2.and there are two main ways that heat is generated: adiabatic heating from extreme plastic deformation and frictional heating between the tool shoulder and workpiece interface. Usually, the temperature distribution reached between 0.6 and 0.8 of the melting points of the base material. The estimation of strain rate and peak temperature (PT) were done based on earlier studies and it is expressed in equation 2.1 and 2.2 (Hoyos et al. 2021; Minhas et al. 2025).

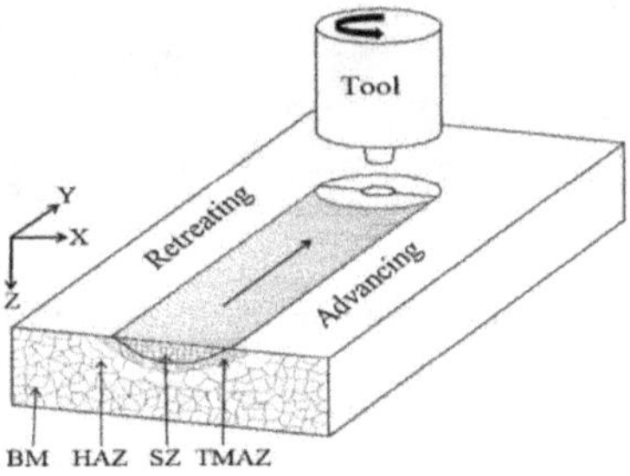

Fig.2.2 Schematic view of FSP (Yadav and Bauri 2012b)

$$\frac{T}{T_m} = 0.62 \times [\frac{N^2}{V} \times 10^{-4}]^{0.06} \qquad 2.1$$

$$\mathcal{E} = \frac{T}{60} \times \ln [\frac{2R_{nugget}}{D_{nugget}} + 1] \qquad 2.2$$

where T, T_m, N, V, $\mathcal{E}$, R_{nugget} and D_{nugget} are melting temperature, peak temperature, tool rotational speed in rad/sec, tool traverse speed in mm/sec, plastic strain rate (s^{-1}), mean radius of dynamically recrystallized zone (DRXed) in mm and the depth of stir zone (SZ), respectively.

The temperature measurement during FSP is being done using Infrared (IR) thermometer (Oliaei et al. 2025). For zinc based aluminium alloys, Guinier Preston (G.P) zones stay stable at temperatures around 150°C. When the temperature reaches above 220°C, eta prime (η′) zones start to form. As it further increase, the G.P. zones disappear, and η′ zones become the main ones that make the material stronger. Between 350°C and 375°C, the eta (η) phase grows quickly. This η phase then dissolves when the temperature reaches 440°C (Canaday et al. 2013; Dixit et al. 2008). So, it is important to maintain the processing temperature below 440^{0}C, to get desired mechanical properties. The temperature range between 350°C and 450°C promotes the coarsening of precipitates, leading to local softening and decreased ductility, indicating how important a role of heat exposure for change in mechanical properties (Mahoney et al. 1998). Study revealed that the PT increases with increase in TRS and reduction in TTS (Sandeep and Natarajan 2022). The relationship between PT and TRS during friction stir welding (FSW) was detailed (Tang et al. 1998).

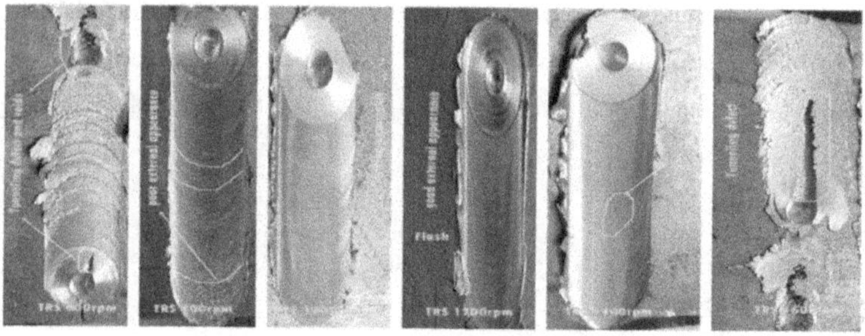

Fig.2.3 Appearance of Friction Stir Processed (FSPed) surface with varying TRS and Fixed TTS (Anandan and Manikandan 2022)

The PT at the weld shows direct relation with TRS, owing to higher frictional heating induced by faster tool rotation, which leads to greater thermal input into the material. A fairly linear trend is seen, indicating that TRS is the primary factor affecting thermal generation during FSW. Since too much heat can cause overaging or softening and too little heat can cause incomplete material bonding, it is crucial to comprehend this relationship in order to manage the quality of the weld. Moderate value of TRS produced PT of less than 300^{0}C, results better material mixing and no formation of defects (Anandan and Manikandan 2022). However, higher TRS produced defects in microstructure. As shown in Fig.2.3, optimum weld quality achieved at 1000 and 1200 rpm, showing smooth surface and no defects. Too low or too high TRS result in defects such as wormholes, flashing or tunneling, owing to improper heat input and material flow. The pseudo-heat index (W) was developed to reflect the effective heat input during the FSP of aluminium alloys, and it is expressed in equation 2.3 (Kim et al. 2017). At maximum W, a defect-free stir zone (SZ) was formed, whereas at minimum W, a tunneling defect was observed.

$$W = TRS^2/TTS \quad (2.3)$$

The selection of machine variables such as TRS and TTS are significantly affected heat input during processing. Too low a TRS and improper mixing of material from AS to RS due to insufficient heat input lead to kissing defects (Zhang et al. 2025). Conversely, TRS that is too high results in crack formation between the matrix and reinforcement material (Suresh et al. 2020). The TTS of the pin significantly influenced the control of heat input, material flow, strain rate, and the reinforcement dispersion into the matrix (Lonavath and Boda 2023). Higher the value of TTS exhibited insufficient heat generation, lead to improper powder distribution, which affects grain refinement (Rana et al. 2016). As TTS decreased, more friction was induced between two surfaces, which enhanced temperature and strain rate generation (Anandan and Manikandan 2022). Authors reported that strain rate is the function of TRS (Kumar et al. 2018).

Table 2.4 TTS and TRS employed during FSP of AA7075

Reference	Matrix	Reinforcement	TRS (rpm)	TTS (mm/min)
(Bahrami et al. 2015)	AA7075-O	SiC	1250	40
(Butola n.d.)	AA7075-T6	B_4C	1000	40
(Jamali and Mirsalehi 2022a)	AA7075-T6	ZrO_2	1400	30-50
(Patle et al. 2019)	AA7075-T6	B_4C & MoS_2	900, 1120 and 1400	25
(Kumar et al. 2022)	AA7075-T6	Al_2SiO_5 and MWCNTs	750 and 1050	20 and 60
(Vasava and Singh 2022)	AA7075-T651	Graphene	710, 1000 & 1400	40, 56 and 80
(Khoshaim et al. 2023)	AA7075	GNPs and BN	1120	30
(Nathan et al. 2023a)	AA7075-T6	rGO	1400	40

The TRS is the most significant factor affecting strain rate. However, at very high TRS, slipping at the tool material interface or excessive heat may decline the effective strain rate. Hence, optimizing TRS is important to balance deformation, heat generation, and material flow for defect-free welds. The increased TRS and decreased TTS caused more heat input, which affects the surface layer thickness, grain size and the distribution of reinforcement (Kurt et al. 2011a). Table 2.4 displays the utilization of TRS and TTS for processing of AA7075 with carbonaceuous and ceramic reinforcements.

Reinforcements employed during FSP: The reinforcements were supplied through pre-existing holes/grooves. The theoretical (V_t) and actual volume fraction (V_a) of reinforcements were calculated based on Equations (2.4) – (2.7) (Ali et al. 2023; Rathee et al. 2017).

V_t of reinforcement = (Area of groove / Tool pin projected area) × 100 (2.4)

Area of groove = Width of groove × Depth of the groove (2.5)

Tool pin projected area = Diameter of pin × Length of pin (2.6)

V_a of reinforcement = (Area of groove / area of surface composite) × 100 (2.7)

Volume percentage (%) of particles played a role to determine the mechanical and tribological properties (Sathiskumar et al. 2013a). The Zenner pining relationship can be used to explain the distribution of particles at grain boundaries and how they hinder grain development, it is expressed in equation 2.8 (Shafiei-Zarghani et al. 2015).

$$d = \frac{4\alpha d_p}{3v_p}$$

(2.8)

Where, d, α, d_p, v_p represents diameter of grain, proportional constant, diameter of particles, volume fraction of particles, respectively.

The equation 2.8 suggests that greater grain refinement can be accomplished by reducing the d_p at a constant v_p of reinforcing particles. Compared to microparticles, nanoparticles have a higher volume because of their larger surface area, which enables more atoms or molecules to be present in a given volume. Thus, according to equation (2.8), grain refining occurred as a result of the increased volume percentage of nanoparticles. The wear resistance of the surface composites was improved by increasing the volume fraction of particles (Barmouz et al. 2011a). The reinforcement particle size is tabulated in Table 2.5.

Table 2.5. Reinforcement particle details

Reference	Reinforcement	Thickness	Diameter	Width	Average particle size
(Shamsipur et al. 2011)	SiC	-	-	-	50 nm
(Vahedi et al. 2020)	Graphene	1-20 nm	-	-	-
(Zang et al. 2021)	GNPs	10 nm	40 μm	-	-
(Jeon et al. 2014)	GO	1 nm	-	-	1-3 μm
(Sharma et al. 2020a)	GNPs	5-10 nm	-	5 μm	-
(Gençer 2024)	Graphite	-	-	-	4.797 μm
(Singh et al. 2024a)	rGO	-	-	-	2-5 μm

A poor surface and mechanical properties of unreinforced aluminium alloy was improved by adding ceramic and carbonaceous reinforcements. Ceramic reinforcements such as alumina (Al_2O_3), silicon carbide (SiC), titanium carbide (TiC), and boron carbide (B_4C) were commonly utilized reinforcing phases in metal matrix composites (MMCs) for improved surface properties (Mondal 2021). In contrast with micro sized particles, nano sized particles showed superior mechanical, tribological and other functional properties, depending on the type and nature of reinforcements (Aghajani Derazkola and Simchi 2020). The improvement in tribological properties of aluminium matrix composites (AMCs) were effectively attainable by incorporating the ceramic particles. Due to their exceptional properties, such as low density, high melting point, high hardness, superior resistance to chemical attack, and high wear resistance, SiC ceramics were preferred in engineering applications (Grinchuk et al. 2021; Mishra et al. 2014; Wang et al. 2021). SiC was frequently used as wear-resistant ceramics in a wide range of tribological applications, such as cylinders, pumps, dies, nozzles, cutting tools, etc. (Zhang et al. 2020).

Ceramics made of silicon and boron carbide have low densities, making them ideal for use as rotating tribo components. SiC is an ideal option for matrix material refinement and it is best suited for applications that demand superior hardness, high thermal stability, excellent wear resistance, enhanced load-bearing capability, and light weight structural components (Kosaraju et al. 2025; Mouria et al. 2024; Veeranjaneyulu et al. 2024). The grain size and wear behaviour of the SiC reinforced composite were significantly influenced by the size of the SiC particles. When compared to macro sized particles, nanosized SiC particles were reduced wear rates and produce finer grains. The tool pin (TP) profile was another key factor influencing the formation of the SiC reinforced surface composite. A straight cylindrical pin profile provided a more uniform distribution of SiC particles and resulted finer grains within the SZ, thereby improved hardness and wear resistance compared to a square tool pin profile (Yang et al. 2024). Carbonaceous materials include graphene (Gr), graphene nanoplatelets (GNPs), graphene oxide (GO) along with ceramic reinforcements were the most appropriate candidate materials for surface modification owing to their unique properties (Bolotin et al. 2008; Lee et al. 2009). Gr has two-dimensional (2D) geometry, structurally consists of a single layer of sp^2-bonded carbon atoms arranged in a tightly packed honeycomb lattice. It has emerged as a highly effective reinforcement for aluminum-based composites owing to its high aspect ratio, lower density (1.06 gm/cc), large surface area (2630 m^2/g), high electron mobility, high thermal conductivity (5000 W/mK), high modulus of elasticity (1 TPa) andstrong interfacial bonding (Papageorgiou et al. 2017; Sharma et al. 2020b).

Mechanical and tribological properties were enhanced after adding Gr to the matrix material (Arab and Marashi 2019; del Valle et al. 2015). A tremendous improvement in hardness and significant decrement in wear loss were achieved on adding Gr to aluminium matrix (Maurya et al. 2016). Owing to its exceptional mechanical properties, GNPs were preferable choice of reinforcement to fabricate nanocomposites, which lead to enhanced mechanical performance. The theoretical strength of graphene nanoplatelets (GNPs) is about 1050 GPa, which is 100 times more than that of steel (Feng et al. 2017). In order to develop the high strength nanocomposites, a variety of liquid and solid-state processing methods were employed (Kumar and Xavior 2014; Rashad et al. 2014). Mechanical properties were improved due intermetallic compound (alumina carbide) formation near GNPs (Zhang et al. 2018). The oxidized graphene layer, known as graphene oxide (GO), has attracted greater interest than Gr because of its comparable qualities to pure Gr and it's more affordable and efficient production. It contains a large amount of oxygen, it may form and develop chemical structures on its surface. GO's high aspect ratio, mechanical properties, and functional groups (epoxy, hydroxyl, and carboxyl), increase bonding, effective load transfer and distribution within aluminium, unlike CNTs and Gr that cluster from van der Waals forces (Obaid and Al-Bermany 2023). Owing to its unique properties, GO is preferd for fabricating composites (Kuzmenko et al. 2008; Marcano et al. 2010). GO is considered as most influencing factor which significantly impact mechanical properties (Aafaq and Jailani 2023). The reduced graphene oxide (rGO) produced defect free microstructure with even distribution within AA7075-T6 matrix owing to phenomena of dynamic recrystallization (DRX) (Singh et al. 2024a). As an extent of DRX, the grains are refined due to GO dispersion within aluminium matrix. Consequently, improved ductility, strength and work hardening rate, than unreinforced alloy (Tahaghoghi et al. 2023). Reduced graphene oxide (rGO) facilitated a uniformly distributed, defect-free microstructure within the AA7075-T6 matrix through dynamic recrystallization (DRX) phenomena, enhancing strength, ductility, and work-hardening rate (Basak et al. 2024; Singh et al. 2024b; Tahaghoghi et al. 2022, 2023). The square groove geometry design ensures proper distribution of nano GO particles, thereby improved mechanical properties (Yadav et al. 2022).

Microstructural analysis of friction stir processed surface: The material undergoes three main zones after FSP namely stir zone (SZ), thermo-mechanically affected zone (TMAZ) and heat affected zone (HAZ). The SZ experienced DRX and intense plastic deformation result in fine, equiaxed grain formation, lead to enhanced strength and hardness. TMAZ, which undergoes partial recrystallization with elongated grains, resulting in improved strength but not as much as the SZ; and the HAZ, where temperature changes without significant deformation, causing grain coarsening and a slight reduction in mechanical properties in comparison to the base material.

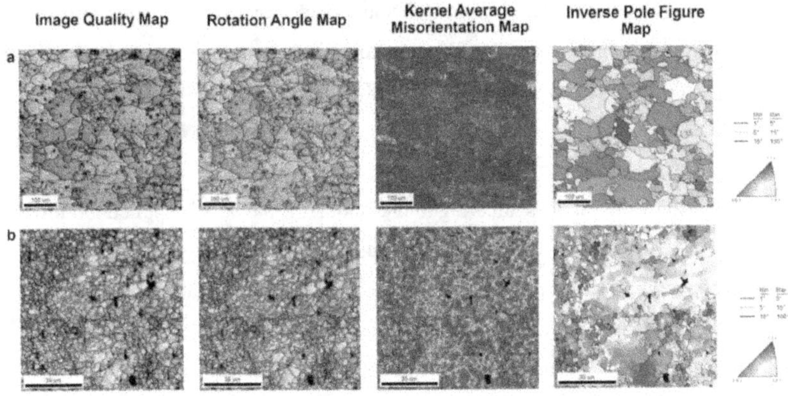

Fig.2.4 Grain structure of (a) AA7075-T6 (b) Surface composite (Khan et al. 2025)

The properties of the fabricated surface composites depends on the distribution of reinforcement particles (RPs) and the microstructure developed in the SZ. The coarse grains are transferred into fine equiaxed grains, owing to reinforcement effect after FSP (Khan et al. 2025). Fig.2.4 (a-b), compares the grain structure of the AA7075-T6 and the fabricated surface composite. As shown in Fig.2.4 (a), the grain size (GS) is non-uniform, ranging from 30 to 60 μm, with a high proportion of high-angle grain boundaries (HAGBs). In contrast, the fabricated surface composite shows (Fig.2.4 b) refined, equiaxed grains with an average GS of 3 μm and an increased existence of low-angle grain boundaries (LAGBs), revealing the effects of temperature, strain, strain rate, and cooling rate, lead to intense plastic deformation and DRX (Liu and Ma 2009). The grain growth was restricted during DRX owing to the higher cooling rate associated with higher TTS, resulting in a refined crystalline structure (Khayyamin et al. 2013). The transformation from HAGBs to LAGBs reflects the influence of FSP in enhancing grain refinement and surface properties. Similarly, the average GS of the nanoparticle reinforced matrix surface was improved by 50 % and 13 % with reference to unreinforced matrix and micro particle reinforced matrix, respectively (Vakili-Azghandi et al. 2024).

The matrix grain size was reduced from 25 μm to 7 μm after adding nano reinforcements of SiC and Gr (Raheja et al. 2020). The dispersion of particles within the matrix is controlled by the optimal values of Tool Traverse Speed (TTS) and Tool Rotational Speed (TRS) (Butola et al. 2021b; Jain et al. 2024; Jamali and Mirsalehi 2022b; More and Jaybhaye 2023). Higher TRS and moderate TTS promoted uniform dispersion of reinforcement particles, leading to refined microstructure and better mechanical performance (Moustafa et al. 2023). The amount of heat used is one of the most crucial factors in FSP. Heat has an impact on the formation of the material's internal structure, which in turn impacts the composite's properties (Maji et al. 2020). Defects may result from the RPs adhering to one another if the heat is applied too low (Mishra and Ma 2005). A TTS of more than 30 mm/min, resulted defects in microstructure, unlike below 30 mm/min (Morales et al. 2018). Overheating causes the grains larger, weakening and softening the material (Karthikeyan et al. 2010). Therefore, to develop strong surface composites, the proper quantity of heat is required. Process variables also affect how RPs mix into the substance. RPs can agglomerate when the TTS is high (Sharifitabar et al. 2011). At higher 'W', significant material softening and enhanced particle distribution were obtained, resulting in a decreased width of the thermo-mechanically affected zone (TMAZ) (Faraji and Asadi 2011). The better microstructural and material properties were achieved with optimized unit stirring (Gangil et al. 2018). Furthermore, the development of the microstructure during FSP depends on machine variables (Kumar and Kumar 2023). TTS has a significant impact on the distribution of SiC particles, surface layer thickness, and particle bonding with matrix substrate. The particle dispersion was uniform at lower TTS, unlike higher traverse speeds (Barmouz et al. 2011b). A well distribution of SiC and its very good bonding with aluminium matrix were attained by moderate TTS (Mishra et al. 2003). A lower TTS produced better surface layer thickness of ceramic reinforced aluminium matrix, unlike higher TTS, which developed weakly bonded surface layer (Kurt et al. 2011b). Another study deals with the influence of TTS on particles distribution of matrix material, which highlights that entangled and aggregated carbon particles were observed when TTS of 100 mm/min and 50 mm/min, respectively. A TTS of 25 mm/min produced good distribution of carbon particles into the matrix material (Morisada et al. 2006).

TTS is considered important on producing weld joints, 25 mm/min along with 1400 rpm produced high quality joints (Ratna Sunil et al. 2015). Another study highlights that TRS and TTS of 1200 rpm and 24 mm/min, respectively, causes localized heating cum severe plastic deformation (SPD) which in turn refines grain structure.(Pradeep et al. 2021). The studies found available on the effect of tool pin profile on material flow behaviour. Fig. 2.5 (a-c) shows different tool pin profiles used (Azizieh et al. 2011). They used shoulder diameter (D) and pin diameter (d) of 18 mm and 6 mm, respectively. The material flow enhanced with the TRS of more than 1000 rpm using threaded TP, further produced onion ring pattern. However, onion rings were not formed owing to lack of heat input below 1000 rpm. The TTS of more than 1000 rpm resulted poor particle distribution and tunneling defect for plain cylindrical TP and fluted cylindrical TP, respectively. Study implies that the D/d between 3 and 3.5, produced onion ring pattern (Saravanan et al. 2016). The TP and TRS played a crucial role for generation of ideal peak temperature, taper cylindrical TP induced greater peak temperature, results defect free microstructure in the SZ (Ratna Sunil et al. 2015). The processing zone temperature significantly drops with increase in TTS. Compare with other TP, cylindrical TP produced maximum peak temperature (Ahmed et al. 2022).

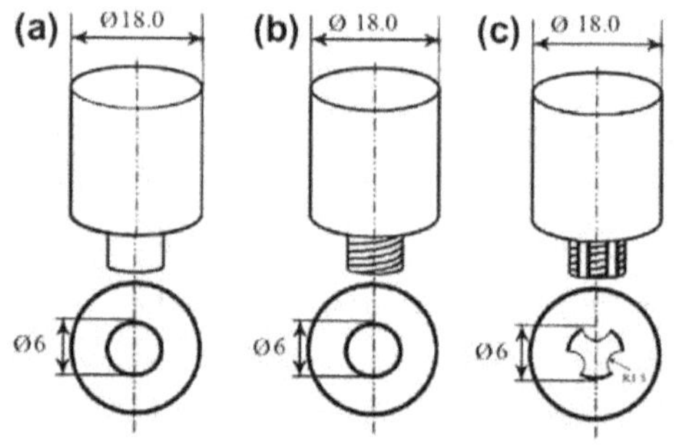

Fig.2.5 Tool pin profiles employed for fabricating surface composites (a) Plain cylindrical (b) Threaded cylindrical (c) Fluted cylindrical (Azizieh et al. 2011)

Mechanical property analysis of friction stir processed surface: Analysing mechanical properties is crucial to comprehending how materials behave under load, stress, and environmental circumstances. AA7075 reinforced with ceramic and carbonaceous particles exhibited improvement in strength, hardness, and wear resistance. This is crucial in applications involving aerospace, marine and automotive components. By analysing such properties, one can predict failure behaviour, and design high strength low density composites that meet safety and performance requirements efficiently and cost-effectively. Tensile strength and hardness were enhanced after adding ceramic particles into a metal matrix. Recrystallization takes place in a stirred area, which is prone to significant frictional heating and severe plastic deformation (SPD), resulting in a refined microstructure (Anil Kumar et al. 2019; Khodabakhshi and Gerlich 2020; Sun and Fujii 2011). The addition of ceramic particles into the aluminium alloy improved ultimate tensile strength (UTS). However, high concentrations of ceramic particles lead to drop in ductility. Developing as-cast lightweight bulk components of metal matrix nano composites with uniform reinforcing distribution and structural integrity would be attracting in order to maintain the ductility (Ferkel and Mordike 2001; Mazahery and Ostadshabani 2011). A threaded TP was employed to process AA7075+SiC surface composite, results good distribution of ceramic particles with enhanced micro hardness (Bahrami et al. 2014). It was also observed that shoulder to pin diameter (D/d) ratio of 3 yielded the maximum tensile strength of dissimilar weld joints made of AA7075-T6 and AA2024-T6 (Saravanan et al. 2016).

The processing variables were significantly affected the mechanical properties, a light weight alloy using taper cylindrical tool with lower TRS and higher TTS resulted defect free joints with exceptional tensile properties (Satheesh et al. 2020). The combination of cylindrical tip tool along with medium TRS and higher TTS produced enhanced ultimate tensile strength (UTS) and microhardness (Butola et al. 2023). The UTS and hardness reflects better when it is processed under low to medium TTS (Butola n.d.; Butola et al. 2020, 2021b). For fabricating AA7075+SiC surface composites, the TTS was identified as most influencing variable with affects microhardness (Mouli and Rao 2021). Hardness showed inverse effect on volume percentage of SiC particles, higher volume percentage exhibited reduced hardness, which lead to ineffective load transfer from matrix to reinforcement particles (Ande et al. 2019). Poor interfacial bonding between AA7075-T6 and SiC particles occurred at high content of SiC particles, lead to reduction in UTS (Duan et al. 2025). It was inferred that moderate TTS, higher TRS and cylindrical pin geometry produced better tensile strength. Defect free stir zone is obtained with uniform distribution of particles under single pass FSPed surface under higher TRS (Mahmoud et al. 2008). The fabricated ceramic reinforced AA7075 surface nanocomposites under L16 orthogonal array by varying TRS, showed inverse relation with mechanical properties (UTS and total elongation) (Butola et al. 2022b). The addition of carbonaceous reinforcement to the surface of heat-treatable alloy improved hardness by a certain weight percentage. Beyond that, a reduction in hardness is observed (Ghazaly et al. 2013). Table 2.6 displays the mechanical properties such as UTS and micro hardness of AA7075 alloys reinforced with various ceramic and carbonaceous particles.

Table 2.6 Mechanical properties of AA7075 reinforced with ceramic and carbonaceous particles

Reference	Matrix material	Reinforcement	UTS (MPa)	Micro hardness (HV)
(Moustafa et al. 2023)	AA7075+AA2024	Zr+WC	253	-
(Patel et al. 2021)	AA7075	Al_2O_3+SiC	243.2	-
(Kumar et al. 2022)	AA7075-T6	Al_2SiO_5 & MWCNTs	-	90
(Ahmed et al. 2014)	AA7075-O	Al_2O_3	-	85

(Bharat and Bose 2024)	AA7075	SiC	290.34	139.42
(Biradar et al. 2025)	AA7075	GNPs	277	135
(Tyagi et al. 2020)	AA7075-T6	SiC+Aloevera	260	120
(Vasava and Singh 2022)	AA7075-T651	Graphene	-	108

The matrix material, primarily AA7075 in various tempered conditions are denoted. The ceramic reinforcements (Al_2O_3 and SiC), carbon-based particles (Gr and GNPs) and hybrid reinforcements (Al_2O_3+SiC and SiC+Aloevera) are also displayed. The type of reinforcement and processing parameters have a substantial impact on the ultimate tensile strength (UTS). The AA7075+SiC composite achieves the maximum UTS of 290.34 MPa, demonstrating efficient load transfer and strong particle matrix bonding. Likewise, Zr+WC and AlO_3+SiC reinforced composites exhibit comparatively high UTS values of 253 MPa and 243.2 MPa, respectively. The microhardness values exhibit a similar pattern, with the maximum hardness seen in composites reinforced with SiC and GNP. Due to their high stiffness and load-bearing capacity, carbonaceous reinforcements enhance hardness and strength, nevertheless, the quality of their dispersion affects their performance. Overall, the data show that carbon-based reinforcements provide notable gains in microhardness and the ability to reduce weight, while SiC and hybrid ceramic reinforcements produce the optimum combination of tensile strength and hardness. The fracture morphology was influenced by reinforcement nature, base material features, and the FSP variables. For instance, the fracture morphology of AA7075-T6 before FSP showed large sized voids without fine dimples indicated coarse grain structure.

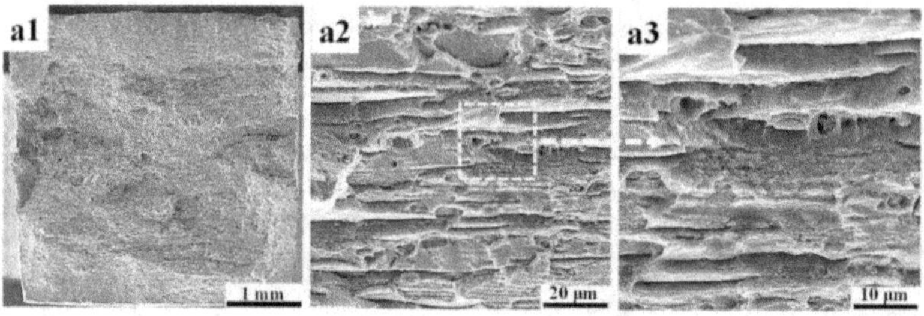

Fig.2.6 (a1-a3) Tensile fracture morphology of AA7075-T6 (Duan et al. 2025)

Appearance of smooth tear ridges in the fracture morphology of AA7075-T6 indicates the brittle mode of failure and it is shown in Fig.2.6 (a1-a3). When ceramic particle was added to AA7075 surface, the mode of fracture was mixed (ductile and brittle) which showed cleavage facets and dimples (spherical and quasi-spherical) (Jamali and Mirsalehi 2022b). The refinement in grain structure was observed after FSP with silicon carbide (SiC) filler addition to AA7075-T6 and its fractured surface showed fine dimples with variable size indicating ductile mode fracture (Deore et al. 2020). The fracture morphology of SiC reinforced surface exhibited finer dimples than unreinforced surface (Hamdollahzadeh et al. 2015).

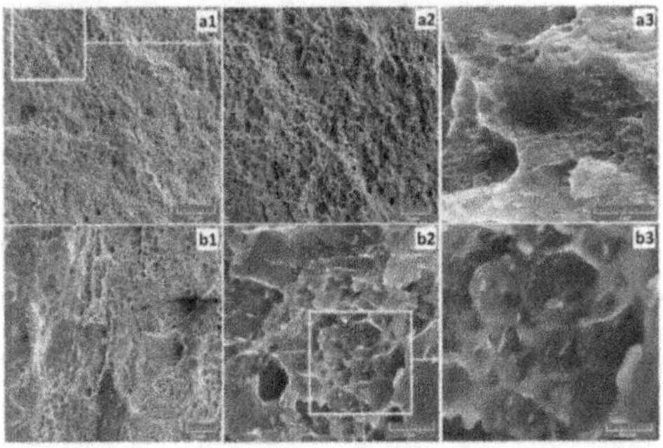

Fig.2.7 Fracture morphology of graphene reinforced aluminium matrix (a1-a3) Lower TRS and Higher TTS (b1-b3) Higher TRS and Lower TTS (Pouraliakbar et al. 2024)

The surfaces of base alloy and FSPed alloy showed voids and dimples, which were signs of ductile fracture, as per fracture morphology. The voids grew finer and shallower, suggesting a shift in the kind of fracture and the emergence of brittle fracture for GO reinforced FSPed alloy (Naghshehkesh et al. 2019). Reduced graphene oxide (rGO) reinforced AA7075 surface after fracture showed deeper dimples indicated pure ductile mode of failure (Singh et al. 2024b). The UTS of fabricated surface nanocomposite showed direct relation with TRS and indirect relation with TTS. The fracture morphology was influenced by TRS and TTS, and is shown in Fig.2.7 (a-b). A considerable amount of ductility prior to fracture is indicated by the uniformly distributed large dimples covering the fracture surface of sample and it is shown in Fig.2.7 (a). The nanocomposite's fracture surface displays a wide variety of dimple sizes, suggested that the matrix and reinforcing graphene were thoroughly mixed. The UTS was reduced owing to minimal ductile fracture mode as evidenced by fewer dimples and it is shown in Fig.2.7 (b).

2.4 SURFACE INTEGRITY OF PROCESSED SURFACE

Surface integrity deals with the collective condition of a component's surface and subsurface details after manufacturing, which encompasses both surface topography (roughness, texture) and surface or subsurface metallurgy (microstructure, hardness, residual stresses, and defects). An interdisciplinary strategy integrating metallurgy, machinability, and mechanical testing is necessary to provide good surface integrity, which especially important in high-stress, safety-critical applications like gears, bearings, turbine blades, and aerospace components, where micro cracks and tensile residual stresses can significantly reduce performance and service life. The surface integrity characteristics are considered as minimum surface integrity, standard surface integrity and extended surface integrity. Surface finish, macrostructure, microstructure, micro hardness, fatigue tests, residual stress, distortion, tensile test, stress rupture, and creep are some of the aspects that go into the evaluation of surface integrity (Field et al. 1972). In order to avoid the superficial flaws of an industrial work piece, protection of surface is most crucial and sensitive. Besides, surface protection is related to service life of the components. Therefore, many researchers paid attention to protect the surface.

Surface roughness is one of the imperative factors for measuring surface integrity. Higher the surface roughness, more will be the surface cracks and less will be the fatigue life (Itoga et al. 2003; Pegues et al. 2018). The surface roughness of larger value produced enhanced pits, which affects the fatigue life (Yao et al. 2017). The machining induced surface defects (micron or nano scale) causes variation in chemical, metallurgical and mechanical effects (Liao et al. 2021). Zuo et al. studied the effects of TTS and TRS on surface topography of AA7075 friction stir welded surface, revealed that TTS was most influential variable that affects the topography of the joint surface. Additionally, higher fractal dimensions result better surface roughness (Zuo et al. 2018). A threaded TP facilitated improved distribution of particles, further improves surface roughness (Rezaee Hajideh et al. 2017). The value of R_a was improved gradually with increase in pass number and it was increased with reduction in TRS (Molla Ramezani et al. 2019a). Lap joint becomes smoother when it is processed between medium to high TRS, on credit to high heat input which improves forming characteristics (Zhou et al. 2017). A moderate TTS resulted better quality of weldment in terms of dimensional accuracy and surface roughness, which fits well for industrial applications (Kumar et al. 2017c).

2.5 CORROSION BEHAVIOUR OF FRICTION STIR PROCESSED SURFACE

Corrosion is a critical challenge in the performance and lifespan of materials used in aggressive environments, such as marine, aerospace and automotive. AA7075-T6, a light weight, high-strength aluminium alloy, is frequently used in these sectors owing to its excellent mechanical properties. However, its corrosion susceptibility limits its widespread use, especially in marine and other harsh environments. This section discusses how the grain size, processing variable and reinforcement nature affect the corrosion behavior of friction stir processed aluminium surface.

Grain size effects on corrosion rate of friction stir processed surface: A critical aspect of FSP is the grain size control, which has a significant effect on the corrosion behavior. The FSW causes microstructural discontinuities in the stir zone, and causes corrosion—hard reinforcing particles in the surface nanocomposite to act as insulators, increasing corrosion resistance (Vijaya Kumar et al. 2015). Traditionally, the correlation between grain size (GS) and corrosion resistance has been well established in earlier reports. In most cases, the grain refinement achieved through FSP tends to improve mechanical properties, which in turn, enhances corrosion resistance by increasing passivation and reduced localized attack. However, certain cases, it may lead to enhance corrosion susceptibility owing to finer grains (Mahmoud 2008; Osório et al. 2005). A finer grain increases the proportion of grain boundaries (GBs), which can enhance corrosion rate owing to the higher energy and chemical activity of GBs (Barbucci et al. 1998; Palumbo et al. 1990, 1991a). The relationship between GS and corrosion rate (CR) is expressed in equation 2.10 (Ralston et al. 2011). From this expression denotes that the corrosion rate gradually enhanced with reduction in grain size.

$$i_{corr} = (A)+(B)\ gs^{-0.5} \qquad (2.10)$$

Where i_{corr}, A, B and gs denotes corrosion current density, environmental constant, material constant and grain size, respectively.

According to an earlier report, decreasing grain size may actually accelerate aluminium's CR. The heat treatable aluminium alloy grain structure was refined after FSP, producing finer grains that were more susceptible to corrosion in hydrochloric acid (Song et al. 2010). The coarse-grained aluminium exhibited reduced corrosion rates, unlike fine grained aluminium, owing to fewer initiation sites for corrosion. Furthermore, the influence of GS is not always straightforward, depending on whether the aluminium grains were equiaxed or columnar, the i_{corr} either enhanced or declined with grain refinement (Vinogradov et al. 1999). As reported across various studies, finer grains reduce the matrix's electrochemical heterogeneity and aid in preventing pit formation. However, excessive refinement of grains showed negative effect on corrosion resistance.

Tool traverse speed effects on corrosion rate of friction stir processed surface: The TTS plays a critical role in controlling the grain size and, consequently, the corrosion behaviour of AA7075-T6 surface composites. The maximum TTS produced minimum heat input, which results coarser grain structure. This coarser structure can increase corrosion resistance by reducing the number of GBs and improving material homogeneity. Additionally, the minimal heat input associated with maximum TTS can help prevent the formation of undesirable microstructural features that could act as sites for corrosion initiation. So, the optimal selection of TTS is essential to prevent corrosion formation. TTS has an inverse effect on the heat input, strain rate, and stirring intensity during the process. Studies highlight that TTS played a role in the distribution of particles within the matrix, regulating the microstructural evolution, improvement of surface properties, and reduction in corrosion rate.(Karthikeyan and Senthil Kumar 2011; Soleymani et al. 2012). Higher tool traverse speeds significantly reduced the corrosion rate, with a greater drop in corrosion current density (i_{corr}) compared to base material, mainly because of grain refinement (Laska et al. 2022; Zainelabdeen et al. 2022). In contrast, moderate TTS allowed sufficient time for mechanical mixing without causing defects (Zang et al. 2021). Carbonaceous nanoparticles processed at medium TTS formed a fine-grained microstructure, increasing corrosion current density and consequently increasing corrosion rate (Aafaq and Jailani 2023). Carefully selecting tool TTS expose surface properties, including corrosion resistance, by refining grains and forming protective layers (Merah et al. 2021; Ralls et al. 2021).

Reinforcement effects on corrosion rate of friction stir processed surface: Surface chemistry of the matrix is being modified on GO addition, improves passivation behavior (Obaid and Al-Bermany 2023). Superhydrophobic silane with GO composite coatings synthesized on the surface of aluminum effectively reduce corrosion rate by blocking corrosive agents (Du et al. 2021). GO coated aluminum alloy surface exhibited a reduction in corrosion current density, indicated an improved corrosion resistance (Candem et al. 2024). The refinement of grains achieved through FSP increased corrosion resistance of the aluminum matrix surface by promoting more uniform, stable passive film formation, lowering surface defects, and preventing localized corrosion onset. Compared with monolithic aluminum alloy, SiC reinforced composites exhibited improved corrosion resistance owing to the formation of protective passivation film (Sarapure et al. 2020). A similar study on nano SiC addition in heat treatable alloy also reported a significant increase in corrosion resistance (Ramadoss et al. 2022). Potentio dynamic polarization (PDP) tests in 3.5% sodium chloride (NaCl) solution confirmed this improvement, attributing it to SiC's cathodic nature, which promoted a protective aluminum oxide (Al_2O_3) layer, declining pitting corrosion (Aperador et al. 2024). Uniform distribution of SiC particles and grain refinement, further improved corrosion resistance (Marode et al. 2023).

2.6 TRIBOLOGY BEHAVIOUR OF FRICTION STIR PROCESSED SURFACE

Tribology is a science of interacting surfaces in relative motion (Menezes et al. 2013). Atoms will be removed from the softer part of material, when it slides against the harder part, and these atoms will typically settle on the harder surface's asperities. As an outcome, interatomic connections form across the interface owing to cold welding at the contact surface. If the sliding process continues, fracture at the junctions may occur, causing the fragments to detach from the asperities. The friction force generated when two surfaces move under an applied force results in shear at these interatomic connections, leading to further separation and wear. A formula for material wear that describes the volume of wear loss (V) of materials as a result of adhesive wear was presented by Archad and it is expressed in equation 2.11 (Archard 1953).

$$V = C\,\frac{PL}{H} \qquad 2.11$$

where V, C, P, L and H, respectively are wear volume loss, wear coefficient, applied load, sliding distance and hardness of the soft contact surface

Tribological performance of metal matrix nanocomposites (MMNCs) was greatly influence by the particle size, a smaller nanoparticle size can lessen the ploughing impact from abrasive wear and reduce potential exfoliation or delamination from adhesive wear (Saba et al. 2019), it can generally introduce good anti-wear performance (Barmouz et al. 2011a). Furthermore, the density and debris distribution of metal matrix nanocomposites (MMNCs) can be indirectly altered by nanophase sizes (Saba et al. 2019), which is also critical for composite tribological behaviour. The anti-wear performance of the material is weakened by larger nanophases, which could yield weaker bonding with the matrix, separation, and the development of micro cracks and delamination pits. The mechanical outcomes and fracture behaviour of the material were affected by strengthening mechanisms such as Orowan strengthening and thermal expansion mismatch strengthening, which were also impacted by the size of the nanophases (Alizadeh et al. 2017; Pan et al. 2019).

Dry sliding tribology behavior analysis under room temperature condition: It refers to the wear that occurs between two solid surfaces in contact, sliding against each other without lubrication, under ambient conditions. In this process, material is gradually removed owing to mechanical interactions such as adhesion, abrasion, or fatigue. The non-availability of lubrication enhances friction, leading to rise in temperatures at the contact surface, which can impact the material's microstructure and wear behaviour. The dry sliding wear testing is most commonly used to evaluate the tribological performance of materials for applications like aerospace and automotive components where resistance to wear is critical. Table 2.7 displays the dry sliding wear testing parameters under room condition during pin-on-disc test for FSPed surface. Lubricants such as liquid and solid are preferred to minimize friction and prevent material degradation owing to wear. However, liquid and grease-based lubricants are not ideal in environments requiring high vacuum, high-speed operations, heavy loads, or extreme temperatures. These lubricants types often fail to perform effectively or have a minimized lifespan in such tribological settings (Tjong 2013).

Solid lubricant coatings can substitute with liquid and grease-based lubricants to minimize friction and wear. These coatings are applied to the surface of material by utilizing physical or chemical vapor deposition (PVD/CVD) methods (Donnet and Erdemir 2004a; b). However, solid lubricant coatings have a minimized lifespan, are difficult to replace, degrade over time due to oxidation, and don't adhere well. Therefore, adding carbon or ceramic based reinforcements to the metal matrix is a promising way to overcome the demerits of both liquid and grease lubricants and solid coatings. Wear resistance of the matrix material was generally improved by adding SiC particles. The wear rate of the AMCs were found to be considerably lower with respect to matrix material and decreased with increasing the content of particle (vol.% or wt.%). The hard ceramic particles resist against destruction action of abrasive and protect the surface, so with enhancing its content, the wear resistance increases. This result aligns with the rule that the materials with greater hardness shows better resistance to wear and abrasive (Das et al. 2001; Ferkel and Mordike 2001; Mazahery and Ostadshabani 2011).

Table 2.7 Dry sliding wear test parameters utilized to test the friction stir processed aluminium surface

Reference	Matrix	Reinforcement	Counter face	Load	Sliding distance	Sliding velocity
(Sudhakar et al. 2015)	AA7075	B_4C+MoS_2	Hardened alloy steel	0.5 kgf	6 km	-
(Rana et al. 2016)	AA7075	B_4C	-	60 N	-	-
(Roy et al. 2019)	AA7075	SiC+CB	-	-	1, 1.5 and 2 km	2, 4 and 6 m/s
(Mehta and Badheka 2019)	AA6061	B_4C	EN31	20 N	2 km	1 m/s
(Moustafa et al. 2020)	AA7075	SiC+BN	Stainless steel	0.3 bar	-	-
(Aydin 2021)	AA7075	ZrO_2	-	5, 10, 20 and 30	-	80, 110 and 140 mm/s
(Jamali and Mirsalehi 2022b)	AA5083	SiC+MoS_2	AISI D3 steel	50 N	60 mm	-

The influence of TTS on wear rate of AA7075+SiC surface composite was reported and it is shown in Fig.2.8 (a-c) (Mouli and Rao 2021). Fig. 2.8 (a) revealed deep scratches owing to poor SiC particle distribution, caused by low TRS and high TTS, on other hand, Fig. 2.8 (b) shows moderate wear with improved particle mixing. The minimal wear and shallow grooves owing to uniform dispersion of SiC, achieved by high TRS and low TTS (Fig.2.8 c). These images reveal that optimal FSP parameters enhanced SiC distribution, enhancing hardness and declining wear. The surface characteristics of FSPed surface nanocomposites are strongly dependent on the type of reinforcement, the amount it contains (vol. % or wt. %), its size/shape, its distribution with matrix material, the size of agglomerates, and its interface with matrix material.

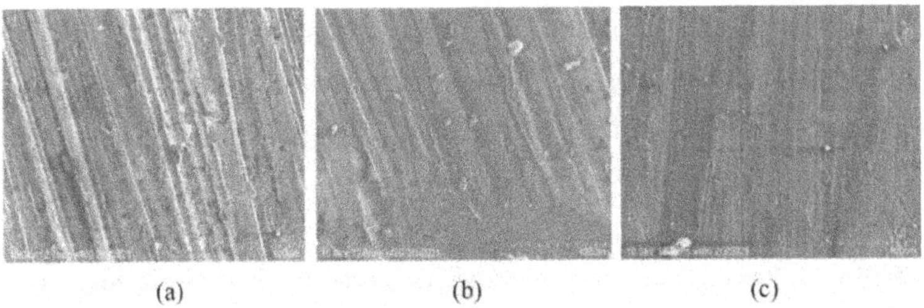

Fig.2.8 SEM photographs showing worn surface of AA7075+SiC composite specimens in experimental runs of wear rate (a) high (b) moderate (c) low (Mouli and Rao 2021)

Adding SiC and boron nitride (BN) to the AA7075 matrix improved hardness and wear resistance compared to the base alloy (Moustafa et al. 2020). The micro-ceramic particle addition to AA7075 produces more wear with an increased coefficient of friction (COF) (Sudhakar et al. 2015). The addition of Tantalum carbide (TaC) and chromium carbide (CrC) ceramic particles on heat-treatable aluminium alloy exhibited a higher percentage improvement in microhardness. However, TaC alone produced the highest wear resistance (Moustafa et al. 2024). Nano ceramic particles (Al_2O_3) and micro carbonaceous particles improve wear performance. The self-lubricating tendency of Gr particles, clubbed with ceramic particles, enhances wear performance and hardness on the aluminium matrix surface due to the formation of a mechanically mixed layer (Ravindranath et al. 2017). Strong interfacial bond attained between aluminium and titanium boride (TiB_2) favours significant improvement in wear properties (Niranjan and Lakshminarayanan 2013). The wear characteristics and wear mechanism of FSPed AA6351+SiC surface nanocomposite were reported (Kumaran et al. 2016; Mehta and Badheka 2019). The wear characteristics of AA7075+SiC/Mg surface fabricated by FSP are detailed (Suresh et al. 2018). The AA7075 reinforced with reduced graphene oxide (rGO) improved wear resistance compared to the base alloy, and the wear mechanism concerned is adhesive (Nathan et al. 2023b). The heat-treatable aluminium alloy is reinforced with 6 wt. % alumina+0.5 wt. % Gr surface produced a lower wear rate (Prakash et al. 2015). The surface of FSPed AA7075+TiC/Graphite yields a minimum wear under 20 N of applied load, and the concerned wear mechanism is fretting (Patil et al. 2021).

Dry sliding tribology behavior analysis under elevated temperature condition: These tests are carried out to analyses the wear behavior of materials under conditions of elevated temperatures, which is crucial for understanding the performance of materials used in environments with extreme thermal loads, such as in automotive or aerospace applications. The purpose of such tests arises from the fact that wear mechanisms can significantly change with temperature, impacting material durability and performance. Hardness showed inverse relationship with wear resistance, whereas COF exhibited direct relationship with temperature. Addition of ceramic particles with AA7075 surface showed better wear resistance than base matrix (Shivaprakash et al. 2023). Wear rate enhances with enhance in temperature and reduces with enhanced in reinforcing agent (Yao-hui et al. 2004). The wear loss showed direct relation with testing temperature, at 200^0C minimum wear loss was achieved, unlike other temperature combinations (Bhaskar 2018). A similar temperature yielded reduction in COF, wear depth and wear rate (Shin et al. 2019). The wear loss showed direct relationship with applied load. The mechanically mixed layer (MML) was formed at 25 N, and beyond this load no MML was seen (Yang et al. 2022).

Tribo-layer was formed for ceramic reinforced aerospace grade matrix material owing to high temperature oxidation, which further reduced coefficient of thermal expansion (CTE) and specific wear rate (SWR) (Siddesh Kumar et al. 2020). SiC reinforced aluminium surface showed minimum COF and SWR, when tested at 150^0C, produced protective glazing layer, and the wear mechanism was changed from delamination (100^0C) to oxidation (150^0C) (Rajaram et al. 2010; Wang and Song 2011). Some studies highlights that slight abrasive wear mechanism was attained when temperature ranges from 25 to 100^0C. When temperature exceeds 150^0C and above, the wear mechanism changed to severe abrasive and fatigue (Zhen et al. 2023). Ceramic particles reinforced AA7075 surface exhibited reduced COF and SWR, when operated under 20 N, 3 m/s and 120^0C of load, sliding velocity and temperature, respectively (Robinson Smart et al. 2020). Reduction in wear loss was obtained owing to the formation of GNPs solid lubricant film in A360 matrix surface and the concerned wear mechanism was oxidation at 150^0C (Martin et al. 2021).

SEM morphologies of wear scars on base material and GNP reinforced surface at room temperature (RT), 150^0 C, and 300^0 C are displayed in Fig.2.9 (a-i) (Martin et al. 2021). The A360+0.25 wt.% GNP shows smoother surfaces owing to solid lubrication, while scratch grooves and delamination patches show abrasive and adhesive wear at RT. Also, 0.25 wt.% of GNP exhibits the least amount of damage; oxide particles provide a protective layer at 150^0 C, lowering wear and friction. The oxide layer ruptures due to matrix softening at 300^0 C, resulting in layered debris and significant wear. The best resistance is provided by 0.25 wt.% GNP, and the overall wear mechanism shifts from abrasive to adhesive (RT) to oxidative (150^0 C) to severe (300^0 C). Table 2.8 displays the summary of high temperature dry sliding wear test parameters for different material and manufacturing combinations.

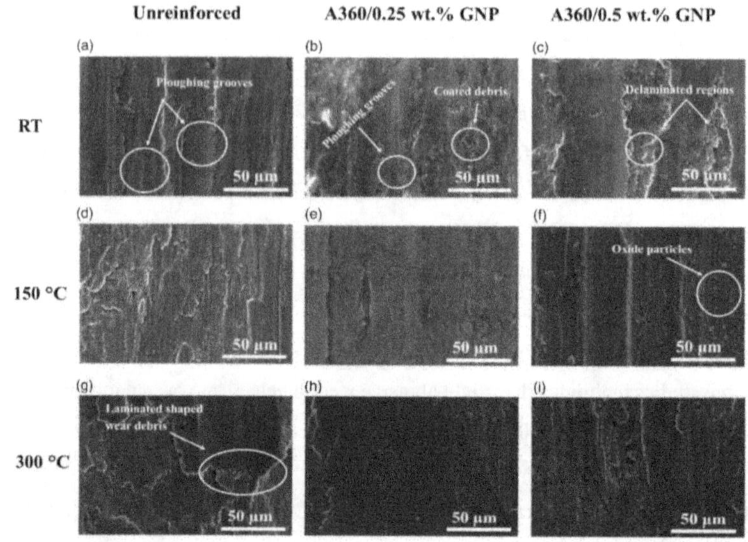

Fig.2.9 SEM images of worn surfaces (a-c) Under room temperature (e-f) at 150^{0}C (g-i) at 300^{0}C (Martin et al. 2021)

Table 2.8 Operating parameters employed under elevated temperature dry sliding wear test of di combinations

Ref	Material combination	Manufacturing route	Wear test o		
			P (N)	SD (m)	V (m/s)
(Siddesh Kumar et al. 2020)	$AA2219+B_4C/MoS_2$	Stir casting	20-60	1500	3.14-5.65
(Wang and Song 2011)	Al+SiC/Alumina	Squeeze casting	30	2500	0.36
(Robinson Smart et al. 2020)	$AA7075+TaC/Si_3N_4/Ti$	Stir casting	20-40	3000	2-4
(Cinca et al. 2018)	Fe-Al	Coating followed by self-propagating high temperature sintering	5	200	0.1
(Martin et al. 2021)	A360+GNPs	Stir casting	2	1000	-
(Shivaprakash et al. 2023)	$AA7075+Si_3N_4$	Stir casting followed by heat treatment	20-50	-	0.41
(Shin et al. 2019)	$AA7075+B_4C$	Liquid pressing	40	200	-
(Rajaram et al. 2010)	Al+Si	Casting	20	2000	1.2
(Yao-hui et al. 2004)	Al+12Si alloy alumina/carbon short fibre	Infiltration	50	942	1.57

2.7 RESEARCH GAP

Based on literature survey of friction stir processed aluminium surface nanocomposites, the following research gaps have been identified

- Studies reported improvement on mechanical and metallurgical properties by varying the tool pin geometry, type of reinforcement and rotational speed. The research on behaviour of surface properties in friction stir processed AA7075 surface nanocomposites by varying tool traverse speed needs attention.
- The influence of tool traverse speed on surface integrity in terms of crystallite size, lattice strain, strength and hardness of AA7075 surface nanocomposites needs attention.
- The addition of specific reinforcement material such as SiC, Gr, GNPs or GO to aluminium alloys has been the subject of numerous investigations. There has not been a consistent study that explicitly compares the specific benefits, such as the high hardness and thermal stability of SiC, the corrosion resistance of GO, and the self-lubricating wear behaviour of Gr and GNPs.
- The influence of tool traverse speed on corrosion and wear behaviour of AA7075 surface nanocomposites needs attention.
- Research paid much attention on dry sliding wear behaviour under room temperature condition of aluminium surface nanocomposites than elevated temperature condition.
- Dry sliding wear behaviour under elevated temperature condition of AA7075 surface reinforced with GO and GNPs need much attention
- Comparison of corroded induced surface roughness in AA7075 surface nanocomposites needs attention.
- Comparison of wear track of the worn out AA7075 surface nanocomposites, an important aspect in tribology is to be explored.
- Utilizing statistical tests such as the Shapiro–Wilk, Levene's and Kruskal–Wallis are significantly enhances the robustness of statistical evaluations. Although these methods are widely recognized, their application in specific research such as mechanical, corrosion and wear studies of surface nanocomposites, needs attention.

2.8 SUMMARY

The following are the summary on review of friction stir processed aluminium surface nanocomposites

- AA7075-T6 alloy advantages, limitations, application requirements and improvement in surface properties are explored.
- The FSP tool design, groove preparation, processing parameters and their influence on heat generation, material flow, dynamic recrystallization, and grain refinement are detailed. The role of tool rotational speed and traverse speed in controlling peak temperature, strain rate, defect formation, and reinforcement dispersion are explored.
- The incorporation of ceramic and carbonaceous reinforcements employed during processing and the details about particle size are explored. The influence of reinforcements on microstructural and mechanical properties are detailed.
- The role of surface protection, surface roughness and surface integrity measures are explored
- Corrosion and tribological performance under room and elevated temperature conditions are explored.
- Key research gaps are identified after literature survey, specifically the influence of tool traverse speed, comparative reinforcement effects, high-temperature tribology, and the use of advanced statistical validation methods.

CHAPTER 3

FABRICATION AND CHARACTERIZATION OF AA7075 SURFACE NANOCOMPOSITES

3.1 INTRODUCTION

This chapter details the fabrication and characterization of AA7075 surface nanocomposites. It beigns with reinforcement particle analysis, followed by AA7075-T6 specimen preparation. The chapter then describes the friction stir processing (FSP) tool with unique profiles, the FSP machine tool with its specifications and variables used, temperature measurement during FSP, and the overall fabrication plan. After fabricating the AA7075 surface nanocomposites, the chapter outlines the equipments and procedures used for micro tensile and micro hardness testing, surface topography evaluation, corrosion and tribology analysis. Characterization techniques such as X-ray Diffraction (XRD), residual stress evaluation, optical microscopy (OM) and Field emission scanning electron microscopy (FESEM) equipped with Energy Dispersive Spectroscopy (FESEM-EDS) are detailed. Statistical methods are included to ensure the robustness of experimental results.

3.2 AA7075-T6 SPECIMEN PREPARATION

Nanoparticles were supplied by different manufactures with purity more than 98 % and varying particle size dimensions. Graphene oxide (GO) and Graphene (Gr) with thicknesses of 0.8 to 2 nm and 10 to 15 nm, respectively, were provided. The Graphene nanoplatelets (GNPs) consisting 10 to 15 layers with a thickness of 10 to 15 nm is being used. The particle size distribution of as received powder samples were analysed using Dynamic Light Scattering (DLS). One milligram (mg) of powder is being taken as a sample size. Using 20 millilitre (ml) of distilled water, sonication is being done for 10 minutes. Then sample was kept in a cuvette, which was placed in the instrument and allowed it for analysis. Particle size was measured using a Nano plus Particle Size Analyser from Particulate Systems. The experimental setup for particle size and zeta potential characterization through Dynamic Light Scattering (DLS) is depicted in Fig.3.1 (a). The instrument measures the zeta potential and particle size distribution in colloidal suspensions.

The sample chamber, where the cuvette is introduced for measurement, is located in the upper part. Figure 3.1 (b) shows the sample chamber lid in close-up, has indication LEDs (for power, temperature, and electric field status) and a button in the middle that opens the lid so that the cuvette holding the sample can be inserted or removed. Fig.3.1 (c) shows the DLS software running on the computer monitor that is attached to the instrument. Fig.3.2 (a–d) illustrates the differential and cumulative intensity plots as a function of particle diameter, while Fig.3.2 (e) summarizes the characteristic particle diameters (D10, D50, and D90). Gr particles are smaller than other particles, which show comparable size characteristics (Fig.3.2 d). Low aggregation is suggested by cumulative intensity and differential intensity curves, which is advantageous for uniform dispersion in composite materials. The standard percentile-based metrics like D (10), D (50) and D (90) are used to describe the particle size distribution.

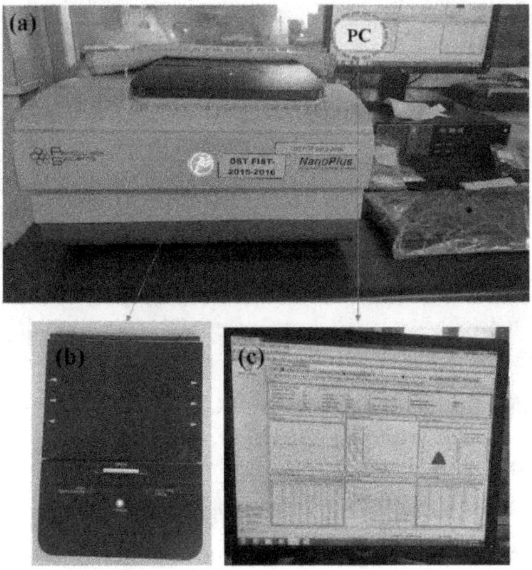

Fig.3.1 Particle size measurement (a) DLS measurements using the Nano plus Particle Size Analyser (b) Cuvette insertion sample chamber (c) DLS software interface showing particle size distribution data.

The values D (10), D (50), and D (90) are frequently employed to characterize the distribution of particle sizes. According to D (10), 10% of the particles are smaller than this size. D (50) represents the median size indicating 50% of the particles are smaller and 50% are larger, this is the size that represents best particles population. As per this, Gr and SiC are strictly classified as a nanoparticle (less than 100 nm), GO and GNPs have sizes that are marginally larger. Ninety percent of the particles are smaller than this value, according to D (90). These numbers aid in assessing the homogeneity and dispersion of particles within a sample. While a larger range denotes size variation (polydispersity), a tight gap between D (10) and D (90) suggests homogenous (monodisperse) particles. The powders under study had median particle sizes ranging from 61.7 nm to 104.4 nm, according to DLS analysis.

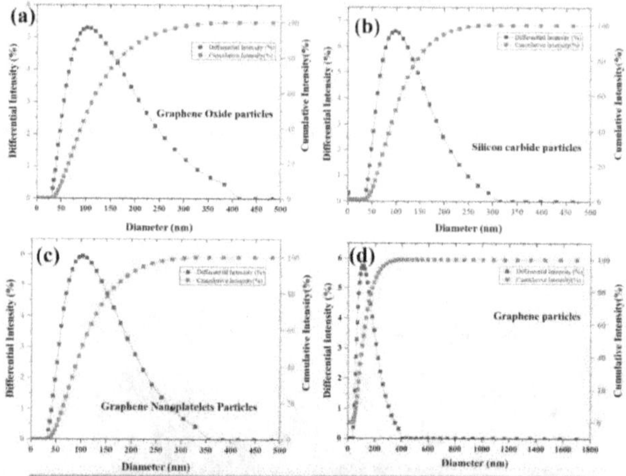

(e)

Powder	D (10%) in nm	D (50%) in nm	D (90%) in nm
GO	52.40	104.40	215.90
SiC	54.70	96.20	173
GNPs	54.60	101.10	192.80
Gr	43.10	61.70	105.30

Fig.3.2 Intensity distribution of particles (a) GO (b) SiC (c) GNPs (d) Particle size distribution (D10, D50, and D90)

GO particles showed a broad size distribution with D50 and D90 of 104.4 nm and 215.9 nm, respectively, revealing significant aggregation and larger particle sizes. Similarly, GNPs exhibited a relatively wide distribution with D50 and D90 are 101.1 nm and 192.8 nm, correspondingly, attributed to the stacking of platelet-like structures. SiC particles showed a narrower distribution with D50 and D90 of 96.2 nm and 173 nm, respectively, consistent with their nature of ceramic. Unlike others, pristine graphene nano particles exhibited the smallest and most uniform size distribution. This narrow distribution suggesting their nanoscale character and reveals a high surface-to-volume (S/V) ratio, which is benefit for applications expecting enhanced thermal, electrical, and interfacial properties. The DLS results confirm that the size of particles and its distribution vary significantly, which directly impacts their suitability for different applications such as reinforcement agents, nanocomposites, coatings and energy storage devices.

Estimation of volume fraction of reinforcement: The mechanical and microstructural characteristics of surface composites made by FSP are greatly influenced by the volume fraction of reinforcements. By adding ceramic or nano scale reinforcements to the surface layer of a metal matrix, enhancement in strength, hardness, wear resistance, and corrosion resistance was acheived. The amount of reinforcement dispersed throughout the stir zone is determined by the volume fraction. An ideal volume fraction guarantees uniform distribution, strong interfacial bonding between reinforcement and a matrix material, and reduced agglomeration. The volume fraction of reinforcements were added to the machined grooves on the base metal surface as per earlier work (Ali et al. 2023; Rathee et al. 2017). V_t and V_a of reinforcement was calculated to be 6.67% and 2.23 %, respectively. Owing to severe plastic deformation during FSP, the stir zone area expanded beyond the tool pin projected area, which in turn, the V_a of reinforcement being less than what was predicted theoretically.

Surface preparation of AA7075-T6 for supplying reinforcements: The AA7075-T6 plate with 150 mm in length, 86 mm in width, and 6 mm in thickness was chosen as matrix material .

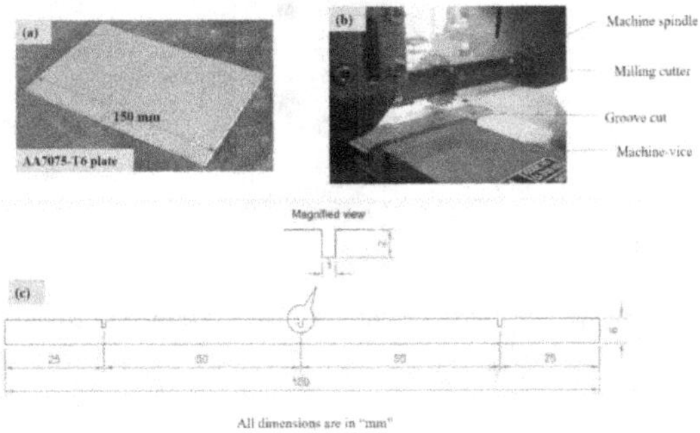

Fig.3.3 Specimen preparation prior to FSP (a) AA7075-T6 alloy (b) Groove being cut by milling machine on AA7075-T6 surface (c) Groove details after machining on AA7075-T6 surface

The chemical composition of AA7075-T6 as follows: Zn 6.2 wt%, Mg 2 wt%, Cu 1.7 wt%, Fe 0.5 wt%, Si 0.4 wt%, Mn 0.1 wt% and Al balance wt%. Its Mechanical properties are Brinell hardness, Young's modulus, UTS, Yield strength and Elongation at break are 150 HB, 71 GPa, 578 MPa, 469 MPa and 11 %, respectively. A horizontal type milling machine was utilized to machine a groove on AA7075-T6 surface. Using a milling cutter, a groove with dimensions of 1 mm in width and 2 mm in depth was machined onto the AA7075-T6 surface to incorporate reinforcements. Fig.3.3 (a) depicts AA7075-T6 plate, processing of groove is shown in Fig.3.3 (b) and details of groove after milling is shown in Fig.3.3 (c).

3.3 FABRICATION OF AA7075 SURFACE NANOCOMPOSITES

An H13 grade tool steel was chosen as tool material due to non-consumable, fits good hardness (60-65 HRC) and found applications in hot working dies (Butola n.d.; Butola et al. 2021a, 2022b, 2023; Tyagi et al. 2020). Fig.3.4 (a-d) shows the FSP tool and its geometry. The cylindrical tool with plain and threaded pin profiles were selected owing to the generation of high value of peak temperature, good metallurgical and mechanical properties (Babu et al. 2020; Dhanesh Babu et al. 2021). Fig.3.4 (a-b) and Fig.3.4 (c-d) shows the plain and threaded cylindrical tool pin profile, respectively. The tool has a shoulder diameter (d_s) of 18 mm, a pin length (L_p) of 5.6 mm, and a pin diameter (d_p) of 6 mm. A tool with threaded pin geometry with a pitch of 2 mm was used.

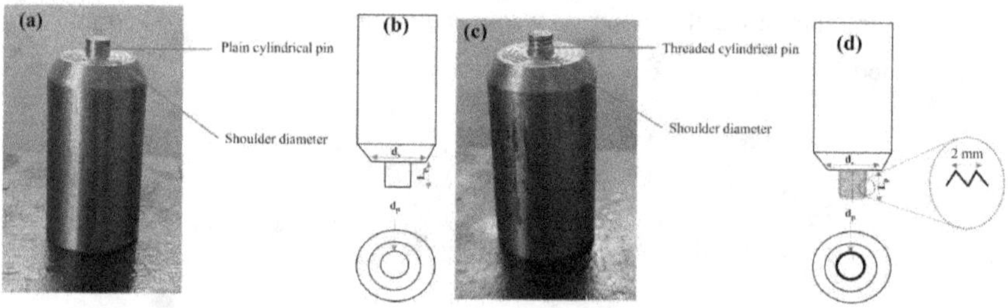

Fig.3.4. FSP tools employed for fabricating AA7075 surface nanocomposites (a-b) Plain cylindrical tool and its pin geometry (c-d) Threaded cylindrical tool and its pin geometry

The fabrication of AA7075 surface nanocomposites were done with a four-axis friction stir welder and it is shown in Fig.3.5 (a). Fig. 3.5 (b) and 3.5 (c) show the FSP of AA7075 surface and the fabricated AA7075 surface nanocomposite sample, respectively. The specifications of friction stir welder are listed in Table 3.1.

Table 3.1 Four axis friction stir welder machine specifications

Variable	Range
Spindle speed	Upto 2000 rpm
Feed	Upto 120 mm/min
Tool tilt angle	Between -3 to $+3^0$
Axial force	Upto 50 KN
Z-axis movement	Upto 900 mm
Maximum allowable work dimensions	300 × 900 mm

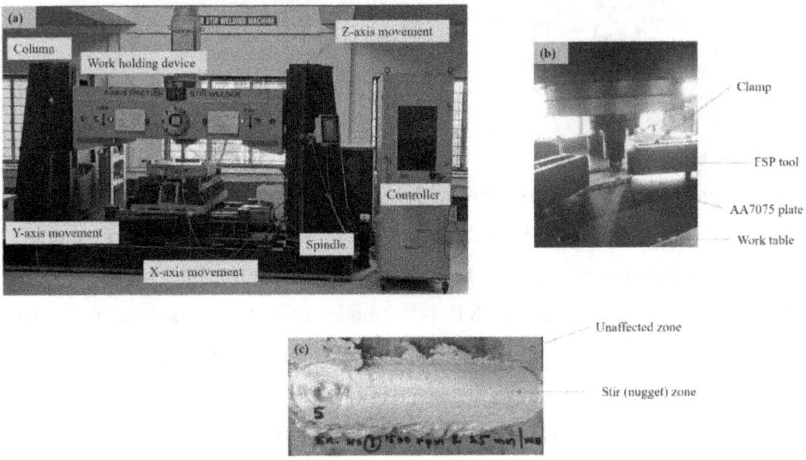

Fig.3.5 Fabrication of AA7075 surface nanocomposite (a) FSP machine tool utilized (b) FSP of AA7075 surface (c) Fabricated AA7075 surface nanocomposite

Improved mechanical properties and uniform distribution of particles within the matrix were obtained when the TRS and TTS were in the ranges of 1000-1600 rpm and 20-40 mm/min, respectively (Butola et al. 2021b, 2023; Moustafa et al. 2020; Satheesh et al. 2020). However, a low TRS produced an insufficient volume of heat generation, leading to weld joint defects (Sevvel et al. 2020). Based on internal trials and previously published optimization studies for AA7075-based surface composites, the TRS and TTS were chosen as 1480-1500 rpm and 20-30 mm/min, respectively. These parameters allowed for uniform reinforcement dispersion while avoiding excessive thermal deterioration or defect formation by offering an appropriate balance between heat input and stirring intensity.

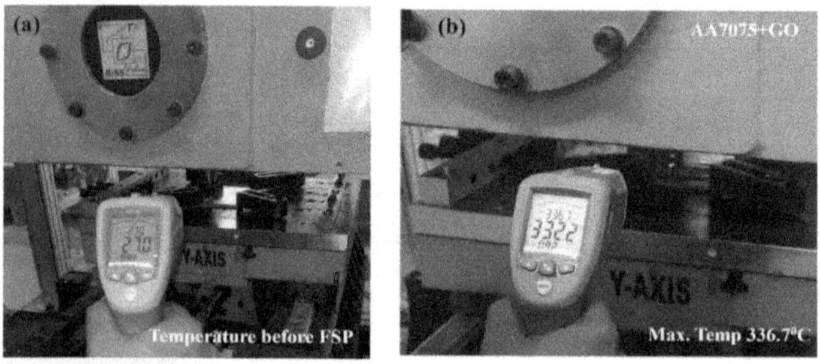

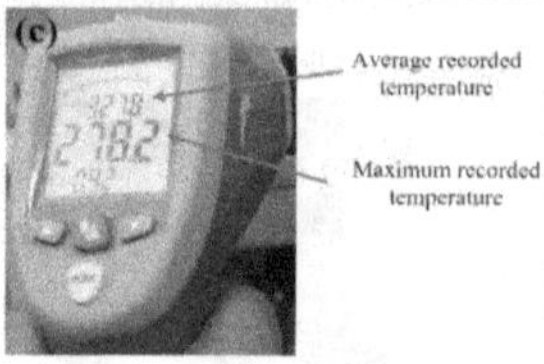

(d)

Reference value of temperature (°C)	Measured value of temperature(°C)	Measurement error (°C)	Measurement uncertainty (°C)
50	48.8	-1.2	1.4
100	98.9	-1.1	1.6
250	247.6	-2.4	1.9
450	448.3	-1.7	3.5

Fig.3.6 Temperature measurement under different condition (a) AA7075-T6 (b) during FSP of AA7075 (c) IR thermometer (d) Calibration of IR thermometer

The measurement of processing temperature was recorded using a non-contact type infra-red (IR) thermometer with DIT-500 model, made by SONEL and it is shown in Fig.3.6 (a-d). Fig.3.6 (a) shows the temperature measurement of base AA7075-T6 alloy and Fig.3.6 (b) shows the temperature measurement during processing of AA7075. IR thermometer with temperature measurements (maximum and average) are depicted in Fig.3.6 (c). Figure 3.7 (d) illustrates the calibration of an IR thermometer following the IW06 procedure for pyrometer calibration. Calibration was carried out at a fixed emissivity value, ensuring SI traceability for accuracy and standardization. An IR thermometer detects the infrared radiation in form of heat energy, emitted by the surface. Later, it transforms this radiation into an electronic signal. This measurement is non-contact, access temperature status from a distance without physically disturbing the hot surface, which is vital during any process involves extensive heat.

Experimental plan for fabrication of AA7075 surface nanocomposites: The fabrication process was structured around two main objectives: first, to study the influence of unique nanoreinforcements (GO, Gr and SiC) at varying TTS (20, 25, and 30 mm/min) and fixed TRS of 1500 rpm using plain cylindrical tool, and second, to study the impact of specific nanoreinforcement combinations (AA7075+GO,

AA7075+SiC, and AA7075+GNPs) processed using threaded cylindrical tool at a TTS of 30 mm/min and TRS of 1480 rpm. In both cases, a tool tilt angle of 0° was maintained. By offering the ideal balance between heat generations and stirring action, the selected TTS range guaranteed efficient dispersion of reinforcements while lowering the risk of thermal deterioration or defect formation. The average of three readings were taken from each sample, and the statistical analysis was done. The samples are identified as displayed in Table 3.2 for convenient representation in this research work.

Table 3.2 Sample identification in this research work

Sample detail	Identification
AA7075+GO processed at 20 mm/min of TTS	GO+20
AA7075+GO processed at 25 mm/min of TTS	GO+25
AA7075+GO processed at 30 mm/min of TTS	GO+30
AA7075+SiC processed at 20 mm/min of TTS	SiC+20
AA7075+SiC processed at 25 mm/min of TTS	SiC+25
AA7075+SiC processed at 30 mm/min of TTS	SiC+30
AA7075+GNPs processed at 20 mm/min of TTS	GNPs+20
AA7075+GNPs processed at 25 mm/min of TTS	GNPs+25
AA7075+GNPs processed at 30 mm/min of TTS	GNPs+30
AA7075+Gr processed at 30 mm/min of TTS	Gr+30

3.4 MATERIAL CHARACTERIZATION TECHNIQUES EMPLOYED

AA7075 surface nanocomposites were subjected to surface integrity, corrosion and tribological analysis. A specific material characterization technique is being done for analysis, involve Optical Microscopy (OM), X-ray diffraction (XRD), Field Emission Scanning Electron Microscopy equipped with Energy Dispersive Spectroscopy (FESEM-EDS), residual stress, a non-contact white light interferometry (WLI), micro tensile and micro hardness. Fig.3.7 (a-c) shows the plane at which the specimen is being cut for carried out surface integrity analysis. The tensile specimen was sliced parallel to the FSP direction, as shown in Fig.3.7 (a) and it was located in the Y-Z plane. The dimensions of the tensile specimen were determined by previous work (Tang et al. 2020). As shown in Fig.3.7 (b), the X-Z plane was used to conduct the microstructure and micro hardness analysis. The X-Z plane was used to collect XRD, FESEM-EDS and residual stress analysis, it is illustrating in Fig.3.7 (c). Fig.3.7

(d) shows different zones present in microstructure of friction stir processed (FSPed) specimen.

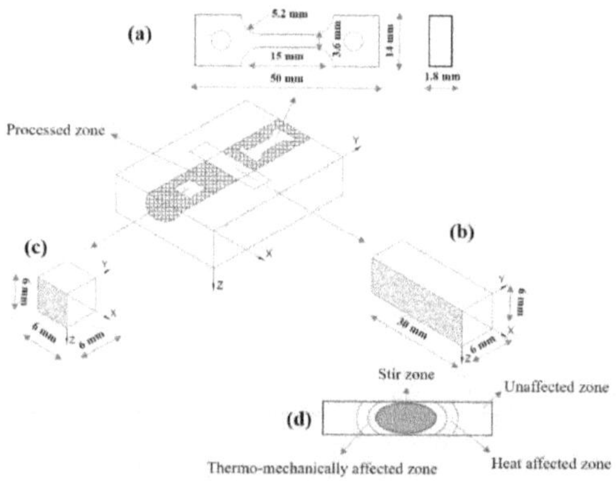

Fig.3.7 Schematic view of the test samples cut from the AA7075 surface nanocomposites (a) Tensile (b) Micro-hardness (c) XRD (d) Different zones

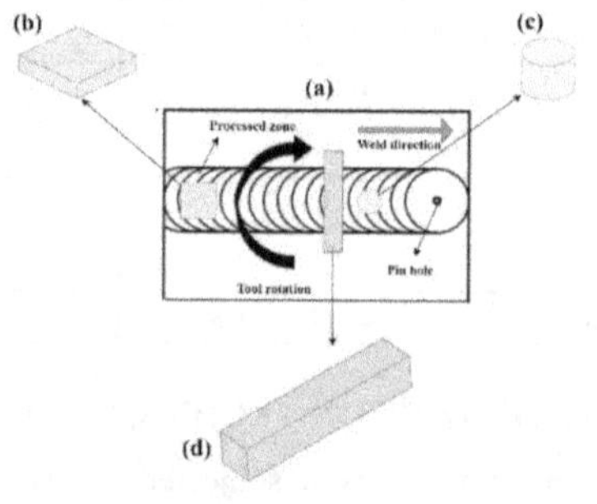

Fig.3.8 Schematic view of the processing of the AA7075 surface nanocomposite wih test samples cutting direction from SZ (a) Direction of weld and tool rotation (b) Corrosion sample (c) Wear sample (d) Microstructural and microhardness sample

Figure 3.8 (a) shows the direction of tool rotation, processed zone, welding direction, specimen for corrosion, wear and microstructural analysis. Fig.3.8 (b) and Fig.3.8 (c) shows the direction at which the specimen is being cut from SZ for carried out electrochemical and tribological analysis, respectively. Fig.3.8 (d) shows the direction at which the microstructure and micro hardness tests were taken. Figure 3.9 (a) shows wire electrical discharge machine (WEDM) is being carried out for preparing the specimen and Fig.3.9 (b) shows the magnified view of machining the fabricated composite

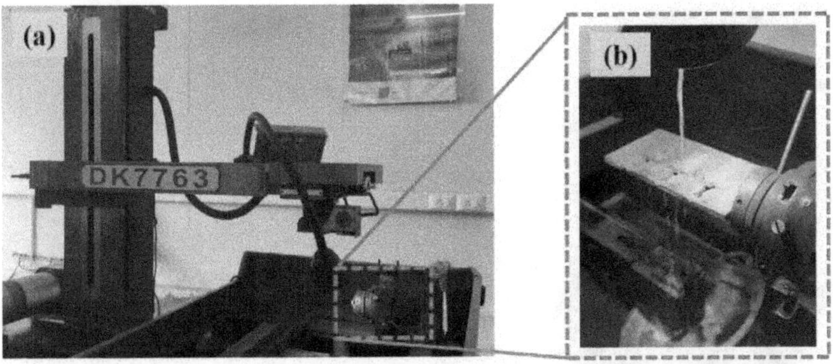

Fig.3.9 Specimen preparation of AA7075 surface nanocomposite (a) WEDM of AA7075 surface nanocomposites (b) Magnified view of machining

XRD examination of AA7075 surface nanocomposites : The AA7075 surface nanocomposites were subjected to X-ray diffraction (XRD) analysis in order to determine the presence of different phases. An XRD sample with dimensions of 6 × 6 mm was examined with the RIGAKU ULTIMA IV diffraction apparatus. Fig.3.10 (a-c) shows XRD equipment with main components and display unit. X-ray tubes generates x-rays and directed towards sample, interact with the crystalline material and they were diffracted as per Bragg's law. The detector records diffracted intensities as a function of 2θ angle, producing a diffraction pattern, which is used to identify phases, crystallite size and lattice strain. The device uses 1.5406Å Kα radiation to scan a range of 5 to 90 degrees at a rate of one degree per minute, with an excitation voltage of 40kV and 30mA of current. The crystallite size (CS) and lattice strain (Ɛ) were calculated using Scherrer equation and Williamson Hall, respectively. CS and Ɛ are expressed in 3.1 and 3.2, respectively (Ikumapayi et al. 2020).

$$CS = \frac{K\lambda}{\beta cos\theta} \quad (3.1)$$

Where CS = crystallite size, K = shape factor (0.94), λ= X-ray wavelength = 1.5406 Å, β= Full-width half maximum (FWHM), θ = Bragg angle.

$$\varepsilon = \frac{\beta cos\theta}{4} \quad (3.2)$$

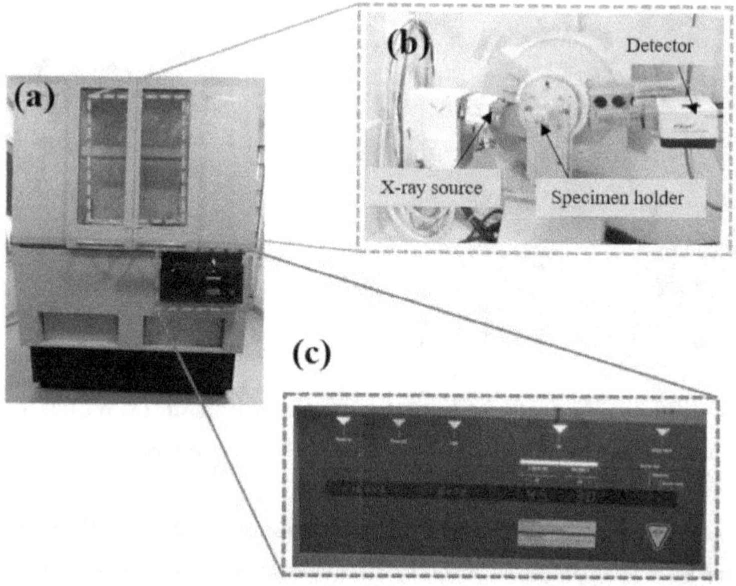

Fig.3.10 XRD equipment employed to detect the interior details of AA7075 surface nanocomposites (a) XRD equipment view (b) Magnified view of components presence (c) Display unit

Residual stress examination of AA7075 surface nanocomposites: The μ-X360 operates based on the XRD technique, which is a non-destructive test to measure residual stress within surface nanocomposites, uses the $\sin^2 \psi$ method. X-rays were directed at the surface of material at different tilt angles (ψ), and the diffracted beams were measured. The position of the diffraction peak shifts depending on the strain within the crystal lattice, allowing the estimation of residual stress using Bragg's law and elastic constants. The Fig.3.11 shows the setup for measuring residual stress. The residual stress analyzer setup consists of a compact X-ray measuring head, mounted on an articulating arm, a specimen mounting for accurate specimen positioning, and a control unit connected to a computer. The system utilizes XRD based on the $\sin^2\psi$ method to measure residual stress. A warning lamp and an emergency stop make sure safety, while specialized software enables real time data acquisition. The specifications of X-ray residual stress analyzer are displayed in Table 3.3.

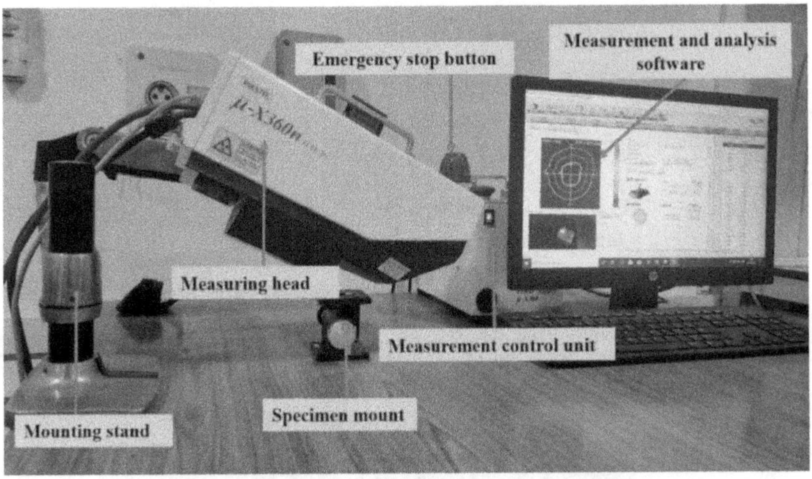

Fig.3.11 Residual stress measurement setup for AA7075 surface nanocomposites

Table 3.3 Specifications of portable X-ray residual stress analyzer

Collimator size	Ø1.0 mm and Ø 2.0 mm for standard and illuminated surface, respectively.
X-ray tube voltage	30kV
X-ray tube current	1.0mA
X-ray incident angle	Single incident angle method
Measurement item	Residual stress, FWHM
Measurement target	Iron (Ferrite) etc.,
Measurement time	Approximately 90 seconds
Power supply	AC 100-240 V, 50-60 Hz, 150 W

Optical micrographic examination of AA7075 surface nanocomposites: The microstructural analysis was carried out on the 30 × 6 × 6 mm cross-sectional specimen surface, which was positioned perpendicular to the FSPed zone. The samples were first cold-mounted, followed by manual polishing. The emery papers of grades 1/0, 2/0, 3/0, and 4/0 were used in consecutive order to polish samples manually. The grade sheets used during manual polishing is depicted in Fig.3.12 (a). The specimens were next cloth polished using a distilled water and Al_2O_3 solution. Fig.3.12 (b) shows cloth polishing machine.

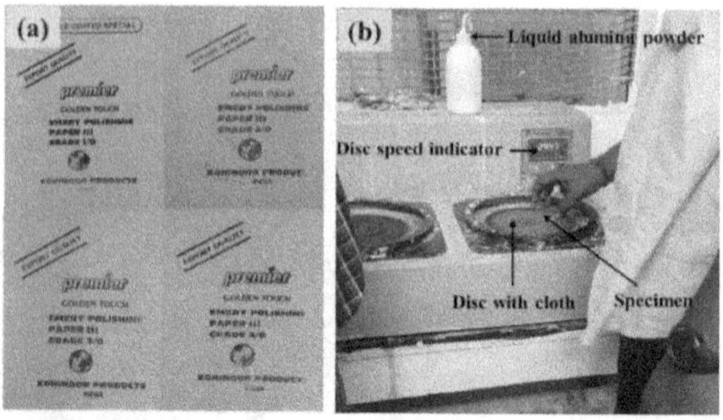

Fig.3.12 Specimen preparation prior to microstructural examination (a) Emery sheets with different grades for manual polishing of FSPed surface (b) Cloth polishing machine utilized for preparing FSPed surface

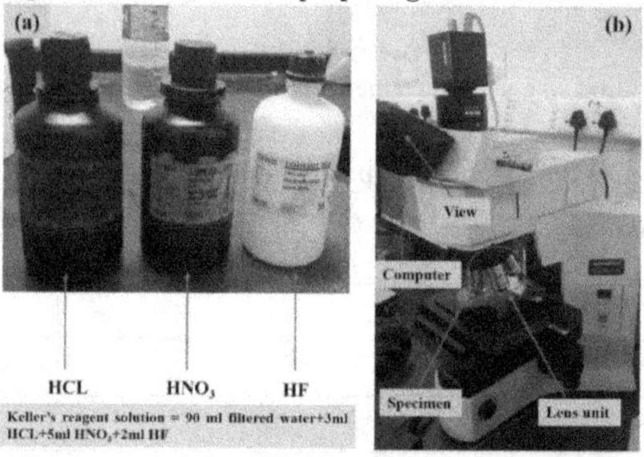

Fig.3.13 Microstructural examination of AA7075 surface nanocomposite (a) Chemicals used for etchant preparation (b) Optical microscopy utilized to view the etched surface nanocomposite samples

Keller's reagent solution contains 90 millilitres of distilled water, 3 millilitre (ml) of hydrochloride (HCL), 5 ml of nitric acid (HNO_3) and 2 ml of hydrofluoric acid (HF). The reagent was then applied on the surface of polished samples to reveal the microstructure. The Fig.3.13 (a) shows the chemicals used for preparing the Keller's reagent. An optical microscope (OM) equipped with an OLYMPUS BX41M-LED model was used to examine the etched samples, as shown in Fig.3.13 (b)

FESEM examination of AA7075 surface nanocomposites: The SEM is a part of extended analysis of nanocomposite surfaces and plays a vital role in failure analysis. In SEM analysis, a component's surface is bombarded with electrons. A primary beam of electrons is generated by an electron cathode for acceleration to the anode for this purpose. This beam of electrons is converged by electromagnetic lenses to a designated object's surface that must be analysed. The so-focused beam of electrons with fine preciseness is scanned line by line across an object that is to be made visible so that this resulting interaction of electrons with the surface is analysed. A series of reactions is induced in a surface when a primary beam of electrons is applied to a sample.

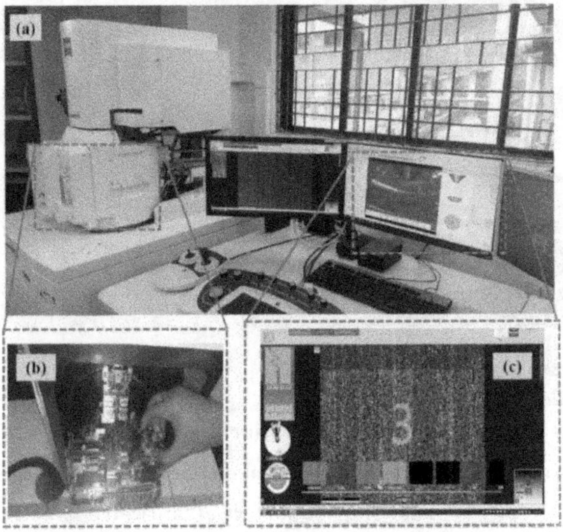

Fig.3.14 Surface morphology examination of surface nanocomposite (a) FESEM equipment (b) Surface nanocomposite sample holder (c) Image scanning mode

The EDS on SEM was employed to analyse the elemental composition of the samples. This equipment also gives a clear view of particles distribution, defect identification, corroded and worn out surface morphologies, in a high-quality resolution, making it highly effective for microstructural investigations in materials research. Fig.3.14 (a-c) shows the FESEM equipment with model and brand of Gemini SEM 300 and Carl Zeiss, respectively. The equipment is capable of examining metallic or non-metallic samples' microstructure at 2000000X magnification with the resolution of 0.8 nm at 15 kV of accelerating voltage. FSPed sample with dimensions of 6 × 6 mm was examined to FESEM-EDS in order to verify the existence of nanoparticles. The FESEM equipment is shown in Fig.3.14 (a). After cleaning of specimens, it is then placed in a holder, as shown in Fig.3.14 (b). Then, the machine is allowed for vacuum purpose for nearly one hour, followed by image scanning using computer software, as shown in Fig.3.14 (c).

Surface topography examination of AA7075 surface nanocomposites: A non-contact white light interferometry (WLI) called the MFT-5000, made from RETEC

Instruments, was used to analyse the surface topography of AA7075 surface nanocomposites. WLI produced high resolution area-based surface data.

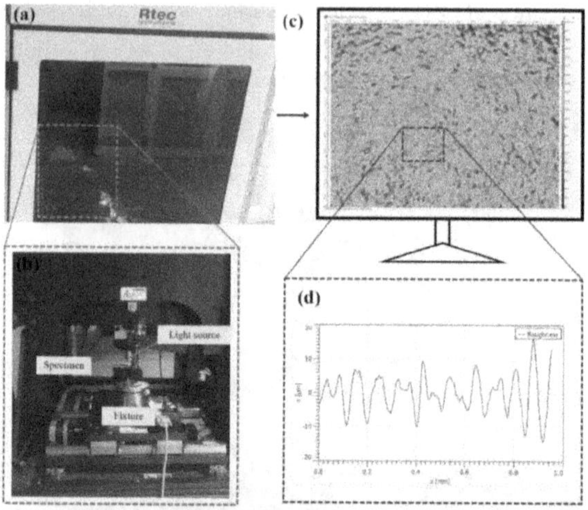

Fig.3.15 Surface topography examination of AA7075 surface nanocomposites (a) A non-contact WLI equipment (b) Magnified view of components present in WLI (c-d) GWYDDION software analysing the surface of fabricated sample

A projected area of 6.86 mm^2 was utilized for measurement purpose from light source. A raw data obtained from the white light source later processed using GWYDDION software. A non-contact WLI interferometry is shown in Fig.3.15 (a-d). Figure 3.15 (a) shows the WLI equipment. The magnified view of components present in the interferometry is shown in Fig.3.15 (b). Fig.3.15 (c) shows the GWYDDION software analysing actual surface after processing and the magnified view of 2D roughness details in the selected area is shown in Fig.3.15 (d).

Ultimate tensile strength examination of AA7075 surface nanocomposites: A TINIUS OLSEN computer-assisted apparatus was used to conduct tensile tests. Fig.3.16 (a) shows the tensile test apparatus. The specimen width, thickness, gauge length and area were measured and the details were feed into the software. Figure 3.16 (b) shows the real time tensile test analysis using horizon software. This gives the clear idea about how the material behaves over the period of time with increasing load. Fig.3.16 (c) shows the fixture which holds the specimen, in which one jaw is fixed and another jaw is movable. The dimensions of the tensile test sample are shown in Fig.3.16 (d), as per previous work.

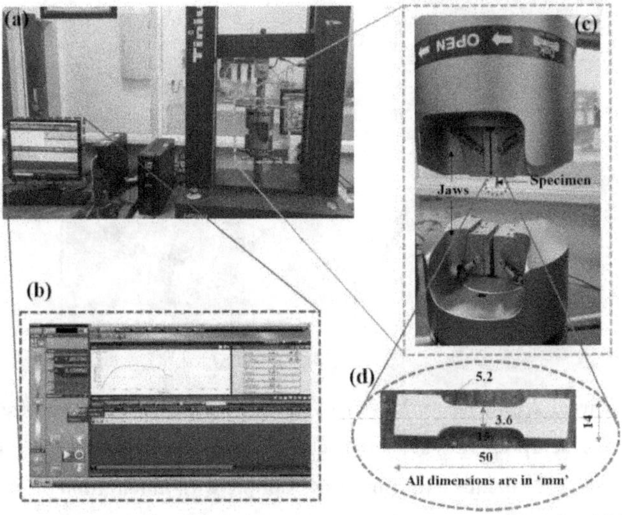

Fig.3.16 Tensile test examination of AA7075 surface nanocomposite (a) Tensile test apparatus (b) Horizon software analysing tensile strength of AA7075 surface nanocomposites (c) Work holding device (d) Dimensions of the tensile sample

Micro hardness examination of AA7075 surface nanocomposites: The AA7075 surface nanocomposites were evaluated using micro hardness equipment with MATSUZAWA, MMT-X model and it is shown in Fig.3.17 (a). The magnified view of indentation is shown in Fig.3.17 (b). For precise and consistent measurements, the amount of time that the maximum load is maintained on the specimen surface is essential (dwell time). A dwell duration of 10-15 seconds is recommended by ASTM E384-22 in order to avoid errors brought on by elastic recovery and to provide adequate plastic deformation, especially in materials that show time-dependent deformation. This guarantees uniformity and comparability of hardness measurements, particularly in the analysis of fine microstructural characteristics. A test load of 300 gram-force (gf) is generally acceptable, as per ISO 6507-1:2023 and ASTM E384-22. This load level provides accurate measurements without sacrificing indentation geometry, making it particularly suited for measuring thin coatings and limited surface zones (ISO 2023). Based on standard testing conditions, 300 gf of load and 15 seconds of dwell time were utilized (Rana et al. 2016). The diagonals (d_1 and d_2) of indentation were measured and the micro-hardness value was displayed. Table 3.9 displays the micro hardness measurements of AA7075 surface nanocomposites.

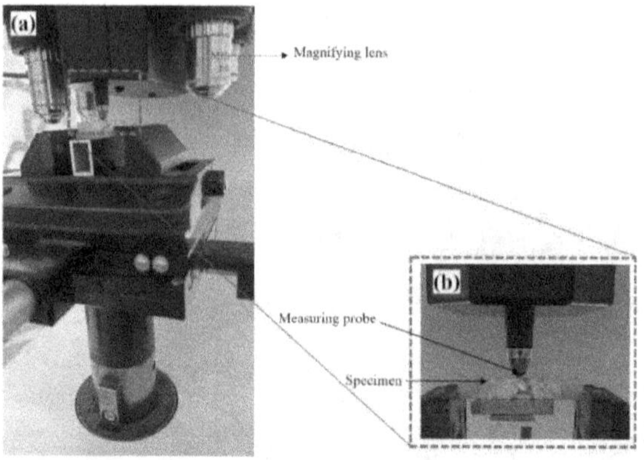

Fig.3.17 Microhardness examination of AA7075 surface nanocomposite (a) Microhardness setup for AA7075 surface nanocomposites(b) Magnified view of AA7075 surface nanocomposite sample with indentation

3.5 POTENTIO DYNAMIC POLARIZATION TEST

The potentio-dynamic polarization test (PDPT) was conducted using 3.5 wt. % NaCl solutions under room temperature as per G102-89 standard (Corro- et al. 2014). Three electrode configuration reference (saturated calomel), counter (graphite), and working (AA7075 surface nanocomposites sample). PDPT apparatus for carrying out electrochemical corrosion test of AA7075 surface nanocomposite samples are shown in Fig.3.18. The electrodes were submerged in a solution contains 3.5 wt. % NaCl. A cathodic potential of -200 mV, an anodic potential of 400 mV, and a scanning speed of 2 mV/s were used as operational settings (Murugabalaji et al. 2024b). The corrosion rate was calculated using the equations (3.3-3.5)

$$\text{The electron equivalent for one gram of an alloy, } Q = \Sigma \frac{n_i f_i}{W_i} \quad (3.3)$$

where, n_i, f_i and W_i are valence, mass fraction and atomic weight of i_{th} element in an alloy, respectively.

$$\text{Equivalent weight (EW)} = \frac{1}{Q} \quad (3.4)$$

$$\text{Corrosion rate (CR)} = K_1 \frac{i_{corr}}{\rho_m} \text{EW} \quad (3.5)$$

*where, K_1 **and** ρ_m are constant (3.27×10^{-3} mm g/µA/cm^2) and density of material in gram per centi-meter cube, respectively. The values for corrosion potential (E_{corr}) and corrosion current density (i_{corr}) are retrieved using a Tafel plot.*

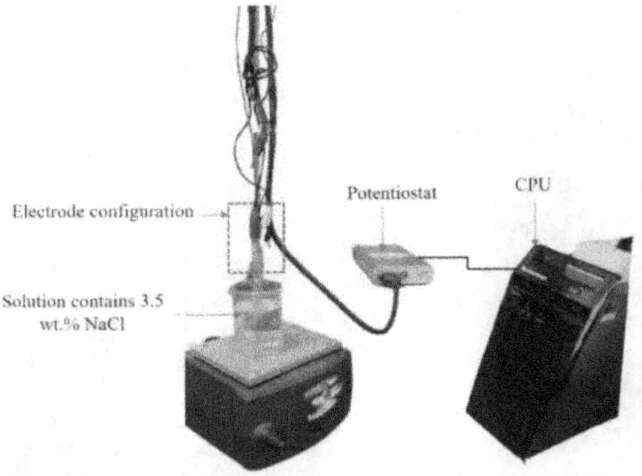

Fig.3.18 PDPT apparatus to carry out corrosion analysis of AA7075 surface nanocomposite

3.6 PIN-ON-DISC TRIBOLOGY TEST

A wear test was conducted on each AA7075 surface nanocomposites as per ASTM G99 standard (International ASTM 2017). The wear test setup, pin and disc details are illustrated in Fig.3.19 (a-c). The DUCOM Pin-on-Disc tribometer setup was utilised to perform the test, as shown in Fig.3.19 (a). Figure 3.19 (b) illustrates a pin with a working specimen for the wear test. The wear test specimen, which was 6 mm in diameter, was cut parallel to the direction of the processing zone and polished using emery paper. The wear test disc is made of EN-31 steel, which possesses 60 HR_C and it is shown in Fig.3.19 (c). The applied load, working temperature, sliding velocity, and sliding speed of 20 N, 150^{0}C, 2 m/s, and 383 rpm, respectively, were used as wear test parameters. To accurately represent the operating thermal conditions of AA7075 based aerospace components, including wing panels, fuselage structures, and fastening systems, a temperature of 150 °C was chosen. Because of aerodynamic heating and their proximity to propulsion systems during flight, these components were frequently exposed to moderate thermal stresses. In these applications, typical service temperatures fall between ambient and around 150–180 °C. Therefore, by simulating the upper range of realistic in-service thermal exposure, where wear and material deterioration can start to develop, the chosen test temperature provides practical relevance.

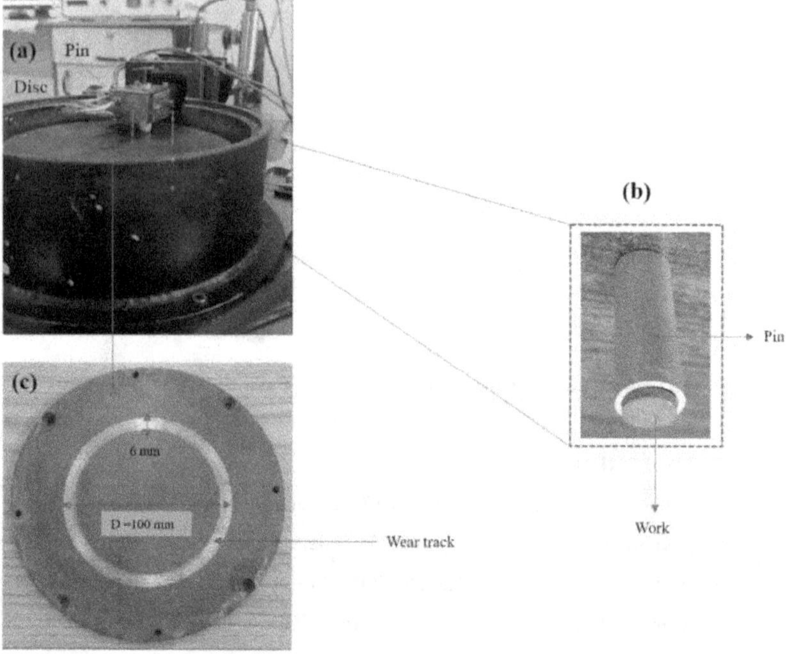

Fig.3.19 (a) Wear test setup showing pin and disc (b) Working sample (pin) of AA7075-T6 surface nanocomposite (c) Disc after wear test showing pin impression, wear track and track diameter

The specific wear rate (SWR) and the coefficient of friction (COF) were the outcomes. The COF was recorded automatically, and the SWR was calculated using equation 3.6, reported previously (Ahmed and Mulapeer 2022). The wear test was conducted for 900 seconds.

$$\text{SWR} = \frac{\delta_m}{(\rho \times t \times F_N \times V_s)} \tag{3.6}$$

Where, δ_m *is mass loss in grams;* ρ *is density of the material in grams/mm*^{3}*;* t *is time in seconds;* F_N *is normal force in Newton and* V_s *is sliding velocity in m/s*

3.7 STATISTICAL ANALYSIS

Statistical tests were conducted, to ensure the robustness of the results. Mean, median, standard deviation (SD), variance (Var) and standard error (SE) were the basic statistical measures, calculated using Microsoft office Excel. Normality assessment of groups was done using Quantile-Qunatile (Q-Q) plots and Shapiro-Wilk test. Q-Q plots were drawn in the relationship between observed quantiles (sorted corrosion rate data) and theoretical quantiles (Z-score). The Z-score is calculated using the equation 3.7.

$$\text{Z-score} = \phi^{-1}\left(\frac{i-0.5}{n}\right) \tag{3.7}$$

Where, ϕ^{-1} inverse of cumulative density function (CDF), 'i' is rank order and 'n' is the number of data points.

Shapiro-Wilk test was conducted on basis of how closely a sample data matches normal distribution (Liang et al. 2009). The Shapiro-Wilk test statistics (W) is calculated using the ratio between weighted sums of the ordered sample values to the total variance of sample and comparing it with 5 % level of significance ($p_{5\%}$). The W is less than the $p_{5\%}$, the data is not normally distributed (H_1) or otherwise the data follows normal distribution (H_0). Box-plot is a visual method to assess variability. It shows median data for each group, Interquartile range (IQR) between top and bottom edge of each box (spread data), and potential outliers (dots outside whiskers). IQR is calculated using the equation (3.8). Outliers are expressed as lower bound (LB) and upper bound (UB). Outliers are calculated using equations (3.9) and (3.10).

$$IQR = Q3-Q1 \tag{3.8}$$

$$LB = Q1-1.5\times IQR \tag{3.9}$$

$$UB = Q3+1.5\times IQR \tag{3.10}$$

Where, Q3 is 75th percentile and Q1 is 25th percentile

Additional tests were used, to guarantee the robustness of variance analysis. Levene's test was performed to test whether the homogeneity of variance exists (Wang et al. 2022). Levene's test statistics (L) is based on the relationship between group and within group variance. By comparing L with $p_{5\%}$, the critical values are extracted. The data is normally distributed, when the critical value is less than $p_{5\%}$ or otherwise not. The Kruskal Wallis non-parametric test assessed median difference among groups, with the test statistics (K) compared to the chi-square distribution at a $p_{5\%}$ (Bai et al. 2024). The null hypothesis is rejected when the critical value is less than K. The K is calculated using the equation (3.11)

$$K = \frac{12}{N(N+1)} \sum_{i=1}^{k} R_i^2/n_i - [3(N+1)] \quad (3.11)$$

Where, N is total number of observations, R_i is the summation of ranks in each group and n is the number of observations in each group.

After careful examination of normality and homogeneity of variability, the analysis of variance (ANOVA) is done. ANOVA was used to find out the significant difference between group means. The sum of squares (SS) was calculated for variables, their interaction, error and total. Let A represents tool traverse speed, B the reinforcement type, and AB is their interaction. The following expressions are used to denote A and B.

$$CF = \sum T^2/N \quad (3.12)$$

$$SS_A = \sum T_i^2/n - CF \quad (3.13)$$

$$SS_B = \sum T_j^2/n - CF \quad (3.14)$$

$$SS_{AB} = \sum T_{ij}^2/n - CF - SS_A - SS_B \quad (3.15)$$

$$SS_{Total} = \sum \sum Y_{ij}^2 - CF \quad (3.16)$$

$$SS_{Error} = SS_{Total} - SS_A - SS_B - SS_{AB} \quad (3.17)$$

Other calculations include:

$$df = n-1 \quad (3.18)$$

$$MS = \frac{SS}{df} \quad (3.19)$$

$$\text{Test statistics} = \frac{MS}{error} \quad (3.20)$$

Where, CF is the correction factor, $\sum T$ is the total sum of all observations, N is the total number of observations across all groups, T_i is the sum of observations in 'i' th treatment group, T_j is the sum of observations in 'j' th treatment group, T_{ij} is the sum of observations for the combination of factor levels 'i' and 'j', n is the number of observations in each group, Y_{ij} is the individual observation, df is the degree of freedom and MS is the mean square.

ANOVA table was framed after calculating above expressions (3.12-3.20). The test statistics was compared with critical values at 5 % significance level and find the significant factor.

3.8 SUMMARY

Fabrication and characterization of AA7075 surface nanocomposites were detailed in this chapter

- Nano reinforcements were successfully embedded into AA7075-T6 alloy after analyzing dynamic light scattering.
- Fabrication of AA7075 surface nanocomposites employed by H13 tool steel with tool rotational speed (1480 and 1500 rpm) and tool traverse speed (20, 25 and 30 mm/min) ensured balanced heat input, uniform reinforcement distribution, and minimized defects.
- Thermal monitoring using a non-contact infrared thermometer ensured controlled thermal cycles during friction stir processing.
- Characterization techniques like optical microscopy, XRD, FESEM-EDS, residual stress analysis and surface topography are used to detail the evaluation of structural integrity and reinforcement dispersion.
- Mechanical performance of nano composite is assessed using micro tensile and micro hardness testing.
- Corrosion resistance was evaluated under 3.5 wt. % NaCl solution using potentiodynamic polarization test following ASTM G102-89 standards.
- Wear behavior was studied using a pin-on-disc tribometer as per ASTM G99, providing reliable tribological performance insights.
- Statistical validation through descriptive, normality, non-parametric and ANOVA analysis ensured robustness and credibility of experimental findings.

CHAPTER 4

SURFACE INTEGRITY OF AA7075 SURFACE NANOCOMPOSITES

4.1 INTRODUCTION

This chapter discuss the effect of different Tool Traverse Speed (TTS) on the surface integrity of AA7075 reinforced with Graphene Oxide (GO), Silicon carbide (SiC) and Graphene nanoplatelets (GNPs) nanoparticles after friction stir processing (FSP). The Crystallite Size (CS), Lattice Strain (Ɛ), Average surface roughness (R_a), Ultimate Tensile Strength (UTS), and Micro hardness (HV) were chosen as surface integrity evaluation criteria. The TTS of 20, 25 and 30 mm/min, were used for fabricating AA7075 surface nanocomposites. The measurement of processing temperature and residual stress at various TTS were analysed. In relation to the TTS, the outcomes derived from surface integrity criteria were contrasted.

4.2 PROCESSING TEMPERATURE ANALYSIS

The temperature measurement using non-contact type infra-red (IR) thermometer are illustrated in Fig.4.1 (a-d). The temperature measurement of as received AA7075 alloy is shown in Fig.4.1 (a) and it is recorded as 27^0 C. The temperature reading was gradually increased with start of experiement, and it is shown in Fig.4.1 (b-d). Table 4.1 displays the influence of TTS and reinforcement type (RT) on temperature profiles, revealed key insights into how TTS and RT affect the maximum and average temperatures during processing. The results suggest that the TTS has a significant impact on the temperature profiles. At 25 mm/min, the maximum temperatures were surge across all combinations of reinforcements. This shows that at moderate TTS, the tool has more time to interact with the material, generating more heat and allowing it to accumulate in the sample surface. Conversely, at 20 mm/min (lower TTS), the heat generated was lower owing to more time for heat dissipation, while at 30 mm/min (higher TTS), the temperatures were lower owing to reduced dwell time, which results in insufficient heat accumulation.

Reinforcing agents also played a key role in determining the temperature profile. Unlike GO and GNPs, SiC produced the highest temperatures, attributed to the enhanced thermal conductivity of SiC, which facilitated better heat transfer during FSP, notably at moderate TTS. GO reinforcement produced lower temperature at higher TTS, indicating that GO may not retain heat as efficiently. GNPs showed a balanced temperature profile, with a moderate enhance in temperature at 25 mm/min, although still minimum than that of SiC. The temperature measurements reveal that moderate TTS are ideal for achieving higher thermal conditions during processing, which can lead to better material flow and homogenous mixing of reinforcements. In contrast, high TTS appeared to result in lower temperatures, potentially limiting the time available for sufficient material plasticization. A previous study revealed that a peak temperature of about 326^{0}C was attained at 1500 rpm of TRS combined with higher TTS, while a TRS of 1000 rpm produced peak temperatures below 300^{0}C, which favored proper material mixing and better mechanical properties (Anandan and Manikandan 2022). These observations suggests the need for careful optimization of TTS based on the material being processed to ensure that adequate heat is generated without compromising the material properties.

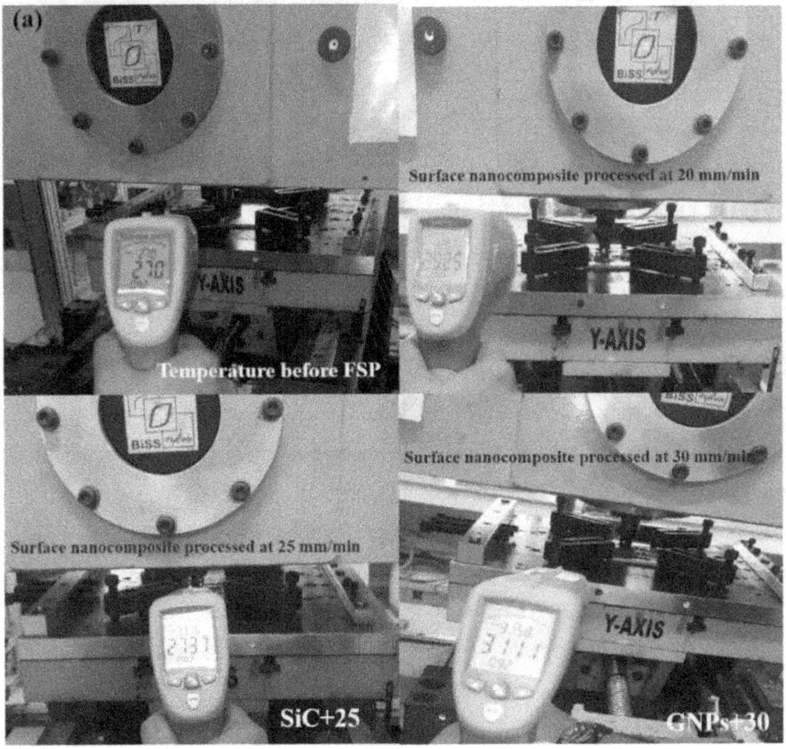

Fig.4.1 Temperature measurement using non-contact type IR thermometry (a) As received AA7075 alloy (b) FSP of AA7075 at 20 mm/min of TTS (c) FSP of AA7075 at 25 mm/min of TTS (d) FSP of AA7075 at 30 mm/min of TTS

Table 4.1 Temperature profile during FSP of AA7075 reinforced with different particles

S.No.	Combination	Readings (^0C) from IR thermometer	
		Maximum temperature	Average temperature
1	GO+20	320.7	292.5
2	GO+25	324.7	303.2
3	GO+30	311.6	279.9
4	SiC+20	323.8	289.9
5	SiC+25	331.6	273.7
6	SiC+30	318.1	309.8
7	GNPs+20	321.2	255.2
8	GNPs+25	328.9	303.8
9	GNPs+30	315.8	311.1
Accuracy = ± 2 to 3^0C; Uncertainity ± 1.4 to 3.5^0C			

4.3 PHASE EVOLUTION, RESIDUAL STRESS AND SURFACE MORPHOLOGY ANALYSIS

This section discuss the AA7075 surface nanocomposite's structural and stress characteristics using X-ray diffraction (XRD), optical microscopy (OM), Field emission scanning electron microscopy (FESEM) and residual stress analysis. OM reveals the microstructure and defects, while FESEM provides high-resolution pictures of surface morphology. XRD identified crystalline phases and detect peak shifts. Residual stress is being analysed using $\sin^2 \psi$ method to understand internal stresses induced during processing. These techniques correlate microstructure, evolution of phases, and distribution of stress, exhibiting a comprehensive understanding of the AA7075 surface nanoocmposite's behavior and performance.

XRD examination of AA7075 surface nanocomposites: The AA7075 alloy detected α-Al peaks and precipitates such as Al_2Cu, $MgZn_2$, η, and Al_2CuMg. The peaks of $MgZn_2$ and Al_2CuMg were reduced after FSP of AA7075 alloy (without reinforcement), suggested partial dissolution. An intermetallic peak ($Al_{23}CuFe_4$) corresponds to precipitation was detected, accompanied by peak shifting confirms lattice distortion and compressive residual stress (Deore et al. 2019). The XRD patterns of fabricated surface nanocomposites under moderate TTS are shown in Fig.4.2 (a-c).

GO, SiC and GNPs combinations detected the prominent peaks associated with α-Al solid solution with h k l indices (1 1 1, 2 0 0, 2 2 0, and 3 1 1) . Owing to the presence of small volume fraction of reinforcements, the corresponding peaks are not detected.

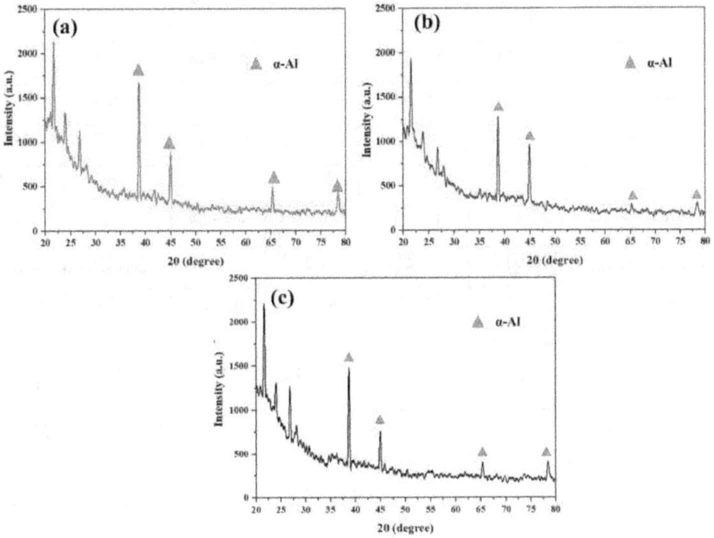

Fig. 4.2 XRD pattern for surface nanocomposites processed at moderate TTS (a) AA7075+GO, (b) AA7075+SiC (c) AA7075+GNPs

The high-intensity peak was analysed to determine the CS, owing to strain-hardening behaviour (Prakash et al. 2018). SiC reinforced AA7075 surface produced smaller CS, unlike others. This provides evidence that Ɛ has an inverse effect on the CS. A minimum CS of 47.162 nm is obtained with a Ɛ of 0.039. This work provides additional evidence for severe plastic deformation resulting from FSP of surface nanocomposites, as seen by the refinement in CS. Full-width half maximum (β) and Bragg angle (θ) were identified by XPERT high score software by using XRD raw data. The mean value, standard deviation (SD) and standard error (SE) of CS is displayed in Table 4.2. The average CS varied significantly between the different materials and processing combinations. The highest average CS of 66.44 nm was observed in the GO+30 combination, revealing that at higher tool traverse speeds, GO facilitate larger grain formation owing to recovery and coarsening of grains. The smaller average CS of 47.18 nm was recorded for GNPs+25, attributed to the effect of GNPs, potentially inhibiting grain growth. The SD and SE values further provide insight into the consistency of these measurements.

Table 4.2 Measured CS of AA7075 surface nanocomposites processed at variable TTS

S.No.	Combination	Readings (nm)			Mean	SD	SE
		R1	R2	R3			
1	GO+20	56.966	58.126	55.452	56.848	1.341	0.774
2	GO+25	47.244	45.188	49.156	47.196	1.984	1.146
3	GO+30	67.233	65.321	66.772	66.442	0.998	0.576
4	SiC+20	56.227	55.711	58.231	56.723	1.331	0.769
5	SiC+25	47.183	47.101	47.202	47.162	0.054	0.031
6	SiC+30	66.611	66.212	66.254	66.359	0.219	0.127
7	GNPs+20	57.751	55.952	56.847	56.850	0.900	0.519
8	GNPs+25	44.786	47.192	49.562	47.180	2.388	1.379
9	GNPs+30	65.872	64.256	69.261	66.463	2.554	1.475

The combination SiC+25 showed minimal variation (SD = 0.054, SE = 0.031), indicating that the CS measurements were highly consistent across the three readings. In contrast, GNPs+30 exhibited more variability (SD = 2.554), which may suggest inconsistent processing or material properties. Table 4.3 displays the estimated Ɛ values for AA7075-T6 surface nanocomposites processed at unique TTS. The data reflects that the Ɛ values shows the variation depending on the material type and the processing speed. GO+30 combination showed the lowest mean Ɛ (0.028), highlighting that at higher tool traverse speeds, the GO surface nanocomposite experiences relatively low deformation. SiC+25 combination showed a higher mean Ɛ of 0.039, indicating that moderate tool traverse speeds may lead to higher lattice distortion in SiC. At TTS of 25 mm/min, the material undergoes intense plastic deformation, promoting grain refinement. The SD and SE reveal that most combinations exhibit minimal variability in Ɛ measurements. The GO+20 combination showed low SD (0.010) and SE (0.006), indicating that the results were consistent and reliable. In contrast, GNPs+25 showed near-zero variation, reflecting very consistent Ɛ measurements across the three readings. In contrast with current findings, earlier research revealed that the Ɛ for the basic AA7075-T6 alloy was less than 0.005, and that the value changed from 0.0050 to 0.0075, following the FSP of AA7075-T6 (Deore et al. 2019). The maximum Ɛ recorded during processing of AA7075+SiC was greater than 0.010. AA7075+carbon filler demonstrated increased Ɛ compared to monolithic alloy and unreinforced AA7075-T6, but decreased with respect to AA7075+SiC.

Table 4.3 Measured Ɛ of AA7075 surface nanocomposites processed at variable TTS

S.No.	Combination	Measurements			Mean	SD	SE
		R1	R2	R3			
1	GO+20	0.022	0.041	0.036	0.033	0.010	0.006
2	GO+25	0.036	0.047	0.034	0.039	0.007	0.004
3	GO+30	0.019	0.036	0.029	0.028	0.009	0.005
4	SiC+20	0.018	0.043	0.038	0.033	0.013	0.008
5	SiC+25	0.027	0.046	0.044	0.039	0.010	0.006
6	SiC+30	0.037	0.023	0.024	0.028	0.008	0.005
7	GNPs+20	0.023	0.039	0.037	0.033	0.009	0.005
8	GNPs+25	0.038	0.038	0.041	0.039	0.002	0.001
9	GNPs+30	0.025	0.026	0.033	0.028	0.004	0.003

Residual stress examination of AA7075 surface nanoocmposites: The residual stress distribution for different combination of materials after FSP at moderate TTS was measured and displayed in Table 4.4. The results showed compressive residual stresses for all fabricated surface nanocomposites, expected to improve material's resistance to fatigue and surface integrity. Under uniform processing condition, SiC reinforcement exhibited the highest compressive residual stress followed by GNPs and GO. This indicates that SiC induces a greater level of compression, stiffness and strengthening effects in the AA7075-T6 surface unlike GO and GNPs. Figure 4.3 (a-c) shows the debye ring patterns, which provide insights into the crystallographic structure and strain behaviour of the FSPed surface. The sharpness and width of those rings are measure of the material's lattice strain and grain size. SiC+25 combination illustrated the narrowest Debye rings, indicative of high lattice compression owing to the significant compressive residual stress. The narrowing of the debye rings indicates a higher degree of crystallinity and strain induced refinement in the surface.

Table 4.4 Measured residual stress of AA7075 surface nanocomposites processed at moderate TTS

S.No.	Combination	Measurements			Mean	SD	SE
		R1	R2	R3			
1	GO+25	−74	−70	−65	−69.667	4.509	2.603
2	SiC+25	−71	−78	−85	−78.000	7.000	4.041
3	GNPs+25	−75	−65	−81	−73.667	8.083	4.667

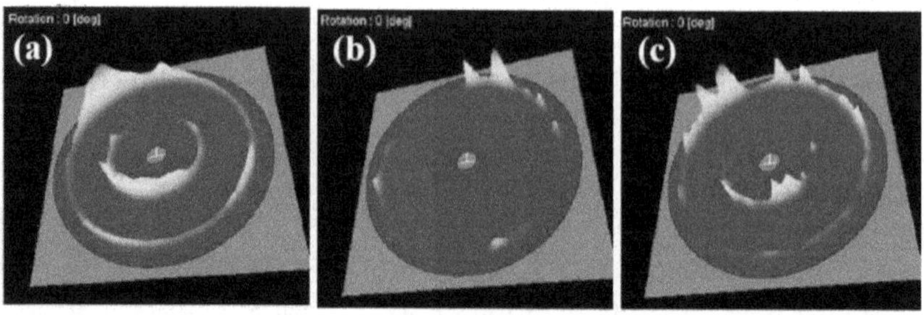

Fig.4.3 Debye ring of AA7075 surface nanocomposites processed at 25 mm/min of TTS (a) AA7075+GO (b) AA7075+SiC (c) AA7075+GNPs

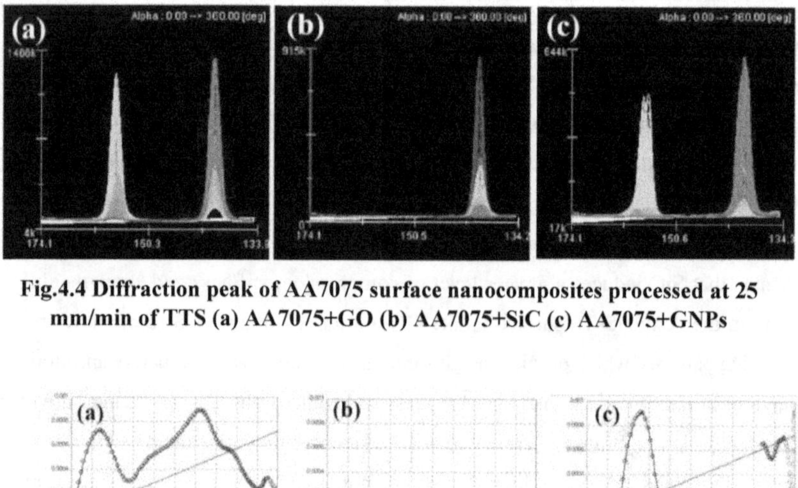

Fig.4.4 Diffraction peak of AA7075 surface nanocomposites processed at 25 mm/min of TTS (a) AA7075+GO (b) AA7075+SiC (c) AA7075+GNPs

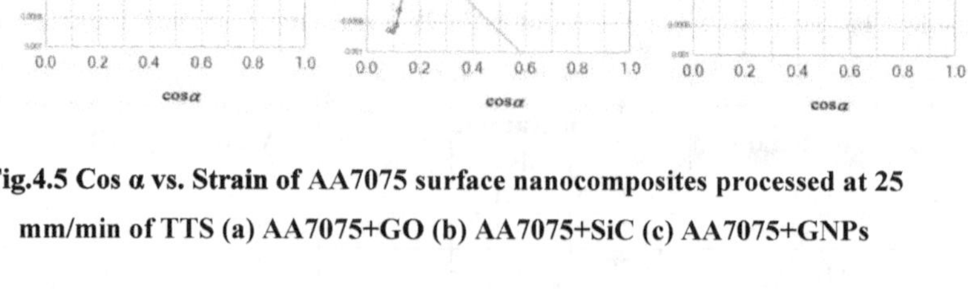

Fig.4.5 Cos α vs. Strain of AA7075 surface nanocomposites processed at 25 mm/min of TTS (a) AA7075+GO (b) AA7075+SiC (c) AA7075+GNPs

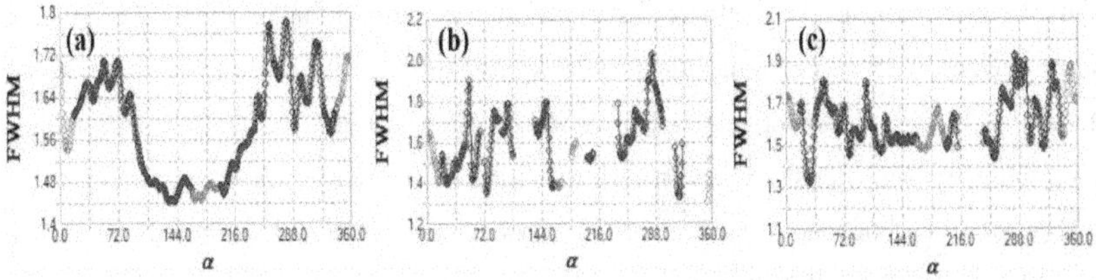

Fig.4.6 FWHM vs. α of AA7075 surface nanocomposites processed at 25 mm/min of TTS (a) AA7075+SiC (b) AA7075+GO (c) AA7075+GNPs

GO+25 showed the lowest compressive residual stress, depicts broader rings, reflecting lower strain. GNPs+25 showed moderately narrow debye rings, suggesting lattice compression in between other combination. Fig.4.4 (a-c) shows the diffraction peaks, reflect the lattice spacing and give evidence of lattice distortion under compressive residual stress.The shape and peak position offer crucial information on how the material undergoes compression during processing. As shown in Fig.4.4 (a), the diffraction peaks were shifted to lower angles, indicating lower lattice compression, consistent with lower residual stress obtained for GO+25. Fig.4.4 (b) shows the diffraction peaks were shifted to higher angles, revealing higher lattice compression for SiC+25, which aligns with higher compressive residual stress (−78 MPa). GNPs+25 shows intermediate performance among those discussed (Fig.4.4 c).

Figure 4.5 (a-c) depicts the relationship between strain and diffraction angle (α). SiC+25 demonstrated higher strain, highlighting that the composite deformed more under compression owing to the enhanced residual stress. GO+25 exhibited the minimized strain, aligns with reduced compressive residual stress. The plot reveals that minimized compression induced deformation, as the composite underwent low elongation. For GNPs+25, it showed moderate strain, consistent with its medium stress. Figure 4.6 (a-c) depicts the relationship between Full Width at Half Maximum (FWHM) and α, which is a measure of CS and defect density. A larger FWHM indicates smaller CS and higher dislocation density, both of which typically connected with larger compressive residual stress and refinement of grains. As shown in Fig.4.6 (a), GO+25 showed the lowest FWHM, suggesting larger CS and lower defect density, unlike SiC+25 combination, which shows largest FWHM, smaller CS and higher defect density. GNPs+25 shows intermediate FWHM values, providing moderate grain refinement and defect density, aligning with its medium compressive residual stress. In contrast with present findings, the residual stress developed for base alloy AA7075-T6 and FSPed AA7075-T6 (without filler) demonstrated low (close to -50 MPa) and high (close to -70 MPa), respectively (Deore et al. 2019). The reinforcement particle was found to have a significant impact on residual stress and lattice strain development. In comparison to carbonaceous particle, SiC reinforced surface composites displayed the largest lattice strains and compressive residual stress (-150 MPa), owing to the grain refinement effect after FSP.

FESEM examination of AA7075 surface nanocomposites: Figure 4.7 (a-f) displays FESEM images along with EDS results for the fabricated surface nanocomposites. The presence of reinforcements like carbon (C) and oxide (O) next to the base aluminium (Al) element is verified, and it shows in Fig.4.7 (a-b, e-f). As shown in Fig.4.7 (a-b), the weight percentage of oxide and carbon combination (79.72%) shows higher than the Al (18.06 %) element, implies dominant presence of graphene oxide. With reference to Fig.4.7 (c-d), the presence of silicon (Si) and C are confirmed and they have more weight percentage (55.13 %) than base Al element. Similarly, the reinforcing agents (80.16 wt. %) shows dominant in AA7075+GNPs surface nanocomposite as shown in Fig.4.7 (e-f).

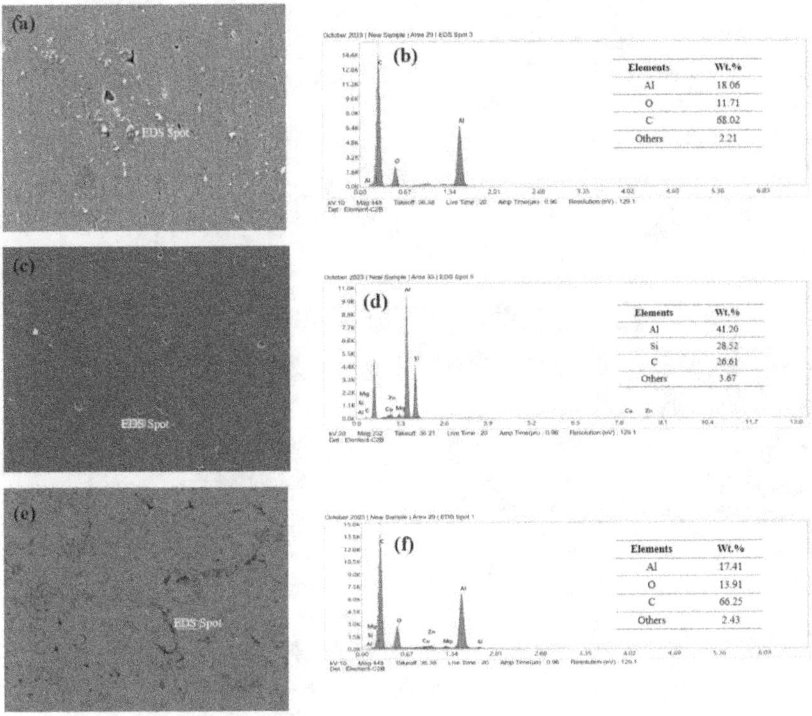

Fig.4.7 FESEM images of surface nanocomposites with EDS spot analysis (a-b) AA7075+GO (c-d) AA7075+SiC and (e-f) AA7075+GNP

OM examination of AA7075 surface nanocomposites: Figure 4.8 (a-f) shows the optical micrographs of the FSPed specimens under different tool traverse speed combinations. The surface nanocomposite produced at tool traverse speeds of 20 and 25 mm/min, respectively, as shown in Fig.4.8 (b-c), has an onion ring structure and no visible internal flaws, such as fractures, cavities, or voids. The onion ring pattern plays a role in enhancing the strength of AA2219-T87 (Kumar R et al. 2019). In comparison with AA7075-T6 base material (Fig.4.8 a), the FSPed zone for three surface nanocomposites (Fig.4.8 b-d) results in fine equiaxed grain structure. The refinement in grains was obtained in SZ due to dynamic recrystallization (DRX) owing to the effect of severe plastic deformation (Satish Kumar et al. 2019). It also shows that grains are independently oriented after FSP.

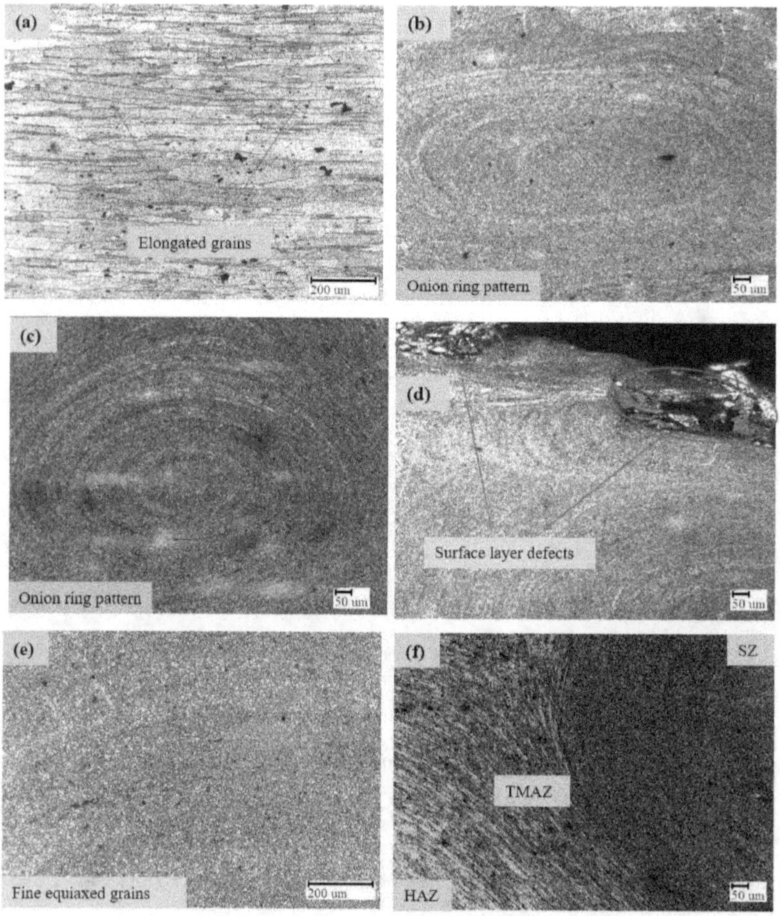

Fig.4.8 Optical micrographs of AA7075-T6 and FSPed AA7075 (a) AA7075-T6 (b-d) AA7075 surface nanocomposite processed at 20 mm/min, 25 mm/min and 30 mm/min of TTS, respectively, (e) Stir zone of AA7075+SiC processed at 25 mm/min, (f) Unique zones identified after FSP

The incorporation of carbonaceous and ceramic reinforcements into AA7075 matrix through FSP produced significant grain refinement compared to the unreinforced alloy, which had limited recrystallization and non-existing of pinning effects (Saxena et al. 2025). The base metal exhibited coarse elongated grains, while SiC and Gr reinforced composites developed fine equiaxed grains owing to Zener pinning and DRX.

The dispersed particles acted as nucleation sites and hindered Grain Boundary (GB) movement, promoting a uniform microstructure (Vasava and Singh 2022). This refinement increased dislocation density and enhanced hardness, unlike unreinforced AA7075. As shown in Fig.4.8 (d), surface layer defects, such as cavities, are formed when a tool traverses at a speed of 30 mm/min. Three separate zones are shown in Fig.4.8 (f), the heat-affected zone (HAZ), the thermo-mechanically impacted zone (TMAZ), and the stir zone (SZ). In contrast to the TMAZ and SZ, the HAZ has longer grains, as shown in Fig. 4.8 (f). The mechanical properties are negatively impacted by these elongated grains. The study found that the fifth-pass friction stir-induced surface nanocomposite had a micro hardness of 134 HV, while elongated grains produced a mean micro hardness of 78 HV (Mehdi and Mishra 2021). This results fine grain structure with 71.80 % improvement in micro hardness with respect to elongated coarse grain structure.

4.4 SURFACE TOPOGRAPHY ANALYSIS

Table 4.5 displays the average R_a readings of AA7075 surface nanocomposites processed at different tool traverse speeds. The data shows significant variation in R_a depending on the material combination and TTS. GO based composites consistently show higher R_a compared to the other materials, the GO+30 combination has the highest mean R_a of 1.92 μm, indicating that higher tool traverse speeds may lead to rougher surfaces in GO composites. This may be attributed to the nature of the GO dispersion and its effect on the material flow during processing. In contrast, SiC based composites are smoother surfaces, with the SiC+25 combination showing the lowest mean R_a value of 1.53 μm. In contrast with present findings, the work reported of R_a on AA7075+SiC showed 2.414 μm at 20 mm/min of TTS, which is higher than 1.65 μm (Molla Ramezani et al. 2019b).

The relatively smooth surface can be linked to the hard and stable nature of SiC, which may resist excessive deformation during the FSP. GNPs also exhibit moderate surface roughness, such as the GNPs+20 combination with a mean R_a of 1.86 μm. The GNPs provide a balance between the smoothness of SiC and the roughness of GO, potentially offering a versatile material for applications requiring tailored surface textures. The SD and SE value s reveal that SiC based composites exhibited consistent surface roughness measurements, as shown in SiC+25 (SD = 0.047, SE = 0.027), while GNPs-based composites like GNPs+20 showed higher variability (SD = 0.248, SE = 0.143). This variation occured due to the dispersion and distribution of the GNPs during the FSP process. These findings suggest that material type and tool traverse speed play significant roles in determining surface roughness. A 3D and 2D surface roughness plot for fabricated surface nanocomposites at 25 mm/min are depicted in Fig.4.9 (a-f).

Table 4.5 Measured surface roughness values of AA7075 surface nanocomposites

S.No	Combination	Measurement (µm)			Mean	SD	SE
		R1	R2	R3			
1	GO+20	1.89	1.78	1.8	1.82	0.059	0.034
2	GO+25	1.82	1.55	1.95	1.77	0.204	0.118
3	GO+30	2.01	1.82	1.93	1.92	0.095	0.055
4	SiC+20	1.59	1.61	1.75	1.65	0.087	0.050
5	SiC+25	1.49	1.51	1.58	1.53	0.047	0.027
6	SiC+30	1.79	1.54	1.92	1.75	0.193	0.112
7	GNPs+20	1.72	1.72	2.15	1.86	0.248	0.143
8	GNPs+25	1.61	1.57	1.69	1.62	0.061	0.035
9	GNPs+30	1.64	1.62	1.91	1.72	0.162	0.094

The TTS of 25 mm/min plays a critical role in shaping the R_a of these materials. As shown in Fig.4.9 (a-b), GO+25 results in the roughest surface, potentially enhancing friction and increasing wear. As shown in Fig.4.9 (c-d), SiC+25 produces a smoother surface, lead to better performance in applications that require minimal R_a. As shown in Fig.4.9 (e-f), GNPs+25 offers a middle ground, providing a balanced surface texture. The surface roughness, in this case, not only depends on the machine variables but also on the material composition and tool profile, highlighting the interplay between machine variables and material characteristics in achieving desired surface properties for specific engineering applications. Similar work reported that the FSPed surface utilized cylindrical tool pin, higher TRS and moderate TTS, recorded lowest R_a (Ahmed et al. 2022). Figure 4.10 illustrates the influence of TTS on the R_a of FSPed surface nanocomposites. The AA7075+SiC surface nanocomposite outperforms the other surface nanocomposites in terms of surface attributes, especially when tool traverse speeds from 20 to 25 mm/min. R_a appears to be declining at a rate of 30 mm/min. All fabricated surface nanocomposites shows superior R_a at moderate TTS.

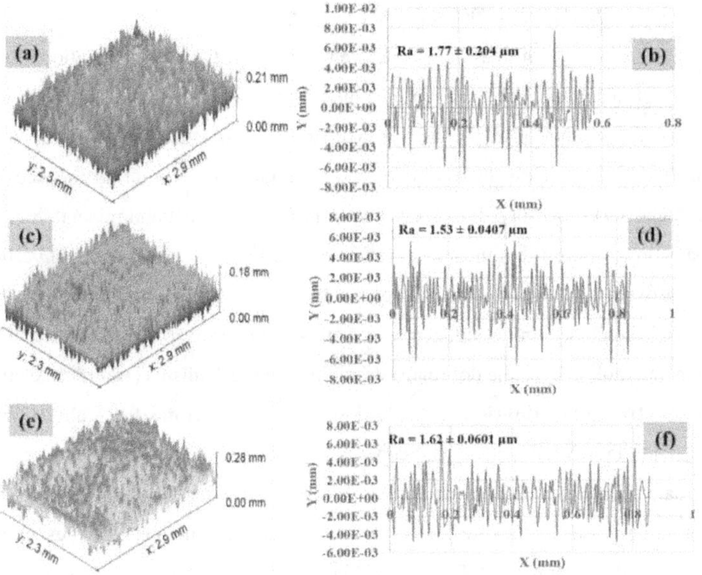

Fig.4.9 3D and 2D surface roughness plot of surface nanocomposites (a-b) GO+25 (c-d) SiC+25 (e-f) GNPs+25

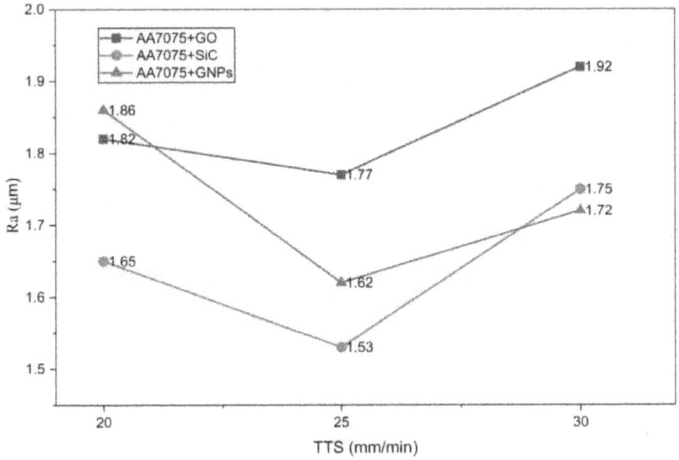

Fig.4.10 Effect of TTS on surface roughness of AA7075 surface nanocomposite

4.5 MECHANICAL PROPERTY ANALYSIS

The strength of materials is affected by the resistance to dislocation movement, a concept also applicable to surface nanocomposites. Lloyd et al (Lloyd 1994) expressed four main strengthening mechanisms in composites namely work hardening, Hall-Petch, mismatch in thermal expansion and Orowan. Kim et al. focused on nano reinforced surface composites, identified as grain refinement and mismatch in thermal expansion were contributing strength (Kim et al. 2006). The reinforcement particles pin grain growth, but beyond a critical volume fraction, they no longer affected grain size, with Hall-Petch becoming independent. At this point, Orowan and mismatch in thermal expansion took over as the dominant strengthening mechanisms (Bodunrin et al. 2015). **Tensile strength evaluation of AA7075 surface nanocomposites:** Table 4.6 displays the UTS of fabricated surface nanocomposite processed at various tool traverse speeds. The data shows significant differences in tensile strength depending on the material combination and processing speed. GO based surface nanocomposites exhibit the lowest tensile strength, with GO+20 having the highest UTS of 330 MPa, while GO+30 shows a significant drop to 242 MPa. This suggests that lower traverse speeds may be more effective for maintaining higher tensile strength in GO composites. SiC surface nanocomposites consistently show high UTS ranging from 347 MPa to 358 MPa, regardless of TTS, highlighting SiC's robustness in maintaining tensile strength under different processing conditions.

Study revealed that the UTS of AA7075+SiC surface composite exhibited UTS of 302 MPa, which is less than AA7075 base alloy, this is due to excessive heat caused during FSP, lead to grain coarsening and dissolution of strengthening precipitates (Kumar and Kumar 2023). Also single pass FSP lead to clustering of SiC particles, which diminishes strengthening effect. GNPs exhibit a more variable response, with GNPs+25 showing the highest UTS (350 MPa), while GNPs+30 has the lowest UTS at 245 MPa, suggesting that higher traverse speeds may negatively affect the material's tensile strength. The SD and SE further highlight the consistency of the measurements. SiC+25, with an SD of 5.292 and SE of 3.055, shows minimal variability in UTS, while GO+30 demonstrates much higher variation in its readings, indicating that the UTS in GO composites is less stable at higher processing speeds. Overall SiC surface nanocomposites demonstrate the highest and most consistent UTS, making them a favourable choice for applications requiring superior tensile strength. For GO and GNPs, lower processing speeds are recommended to achieve higher tensile strength. In contrast, unreinforced FSP of AA7075 demonstrated 205 MPa of UTS, which were processed at higher TTS. Present findings shows that UTS value of more than 205 MPa, indicates the influence of reinforcing effect.

Table 4.6 UTS of AA7075 surface nanocomposites

S.No	Combination	Measurement (MPa)			Mean	SD	SE
		R1	R2	R3			
1	GO+20	315	325	350	330	18.028	10.408
2	GO+25	347	340	348	345	4.359	2.517
3	GO+30	216	230	280	242	33.645	19.425
4	SiC+20	349	342	350	347	4.359	2.517
5	SiC+25	352	360	362	358	5.292	3.055
6	SiC+30	253	240	260	251	10.149	5.859
7	GNPs+20	344	328	342	338	8.718	5.033
8	GNPs+25	350	345	355	350	5.000	2.887
9	GNPs+30	255	220	260	245	21.794	12.583

Figure 4.11 (a-c) shows the tensile fracture pictures of surface nanocomposites that were processed at a tool traverse speed of 25 mm/min. As depicted in Fig.4.11 (a), the AA7075+GO surface nanocomposite showed many equiaxed tiny dimples, indicating a ductile mode of fracture. The greatest percentage of elongation before fracture is indicated by an extremely fine microstructure, which is indicated by a significant number of microscopic dimples (Jha et al. 2023). Conversely, the AA7075+SiC surface nanocomposite, as shown in Fig.4.11 (b), shows large dimples surrounding the cleavage facets that different in size, confirming the combination of dual mode of fracture (Kaviyarasan et al. 2023). However, the dimple distribution shows minor than the formation of cleavage facets. The inclusion of SiC particles to AA7075-T6 matrix results the formation of more cleavage patterns which leads to increment in mechanical properties such as micro hardness and UTS. The fracture pattern appears in AA7075+SiC surface nanocomposite shows similar to fluvial morphology (Cui et al. 2020). While comparing to other surface nanocomposites, AA7075+GNPs shows deeper and honey comb dimples as shown in Fig.4.11 (c), attributed by ductile fracture mode. Also, the dimples are not evenly distributed and it also shows the variation in morphology.

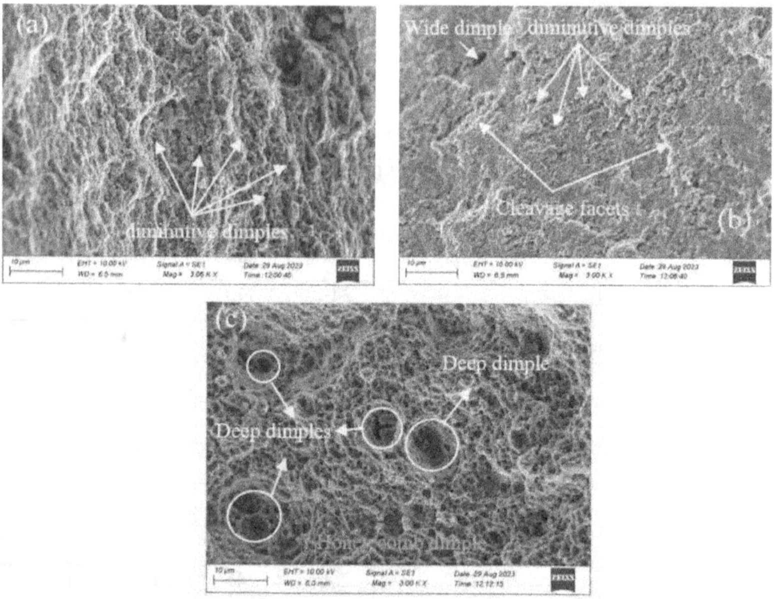

Fig.4.11 Tensile fracture morphology of surface nanocomposite processed at 25 mm/min (a) AA7075+GO (b) AA7075+SiC (c) AA7075+GNPs

The mixed characteristics of dimples presented in GNPs reinforced nanoparticles deficit mechanical properties. As per past works (Tandon et al. 2025), the fractography displayed that AA7075 fractures with large dimples and shear facets, indicative of brittle mode of failure. FSPed AA7075 (unreinforced) at 25 mm/min of TTS, surface showed finer dimples and no shear facets, suggesting a more ductile fracture mechanism (Kumar et al. 2017b). The microstructural variation was consistent with refinement in grain, dissolution of coarse precipitates, and enhanced strain accommodation, leading to improved ductility and stable deformation behaviour.

Micro hardness evaluation of AA7075 surface nanocomposites: Table 4.7 displays the micro hardness measurements of fabricated surface nanocomposites. The data shows notable differences in micro hardness depending on the material combination and TTS. SiC based surface nanocomposites show the highest and most consistent micro hardness, with SiC+25 achieving a mean of 139 HV, which is the highest across all samples. This reveals that SiC composites are particularly resilient to indentation, making them suitable for applications requiring superior hardness.

Table 4.7 Measured Micro hardness at SZ of AA7075 surface nanocomposites

S.No	Combination	Measurement (HV)			Mean	SD	SE
		R1	R2	R3			
1	GO+20	128	120	121	123	4.359	2.517
2	GO+25	137	128	125	130	6.245	3.606
3	GO+30	117	118	110	115	4.359	2.517
4	SiC+20	124	128	129	127	2.646	1.528
5	SiC+25	137	138	142	139	2.646	1.528
6	SiC+30	120	118	122	120	2.000	1.155
7	GNPs+20	120	124	131	125	5.568	3.215
8	GNPs+25	134	129	136	133	3.606	2.082
9	GNPs+30	126	110	115	117	8.185	4.726

For all cases, a 30 mm/min TTS produced lower hardness value. Similar work reported that AA7075+SiC surface nanocomposite processed at 1500 rpm of TRS and 30 mm/min of TTS, resulted a hardness of about 115.71 HV (Molla Ramezani et al. 2019b). In contrast, GO-based surface nanocomposites shows more variability in micro hardness, with GO+25 (130 HV) exhibiting higher hardness than GO+20 (123 HV), but GO+30 (115 HV) showing a significant drop. This indicates that the higher tool traverse speeds may negatively impact the microhardness of GO composites. GNPs based surface nanocomposites show moderate microhardness values, with GNPs+25 showing the highest mean hardness (133 HV), while GNPs+30 has a lower value of 117 HV, indicating that the processing speed affects the microhardness in GNP-based surface nanocomposites as well. The SD and SE highlight the consistency of the measurements. SiC+25 demonstrates minimal variability (SD = 2.646, SE = 1.528), while GNPs+30 shows higher variability (SD = 8.185, SE = 4.726), suggesting that the dispersion of nanoparticles in the matrix might influence the hardness consistency. Overall, SiC composites provide the highest and most consistent micro hardness, making them ideal for applications requires critical wear resistance . For GO and GNPs, optimizing processing parameters could help achieve better and more consistent hardness values. Studies demonstrates the impact of precipitate ($MgZn_2$) on the FSP process. For Instance, the tiny details start to dissolve, get coarser, or reverse when the precipitates are exposed to intense heat brought on by severe plasticity as a result hardness or strength significantly decreases. With reference to Fig.4.12, the surface nanocomposite has shown greater hardness at the SZ owing to the existence of nanoparticles and it started begun to degrade away from it. Similarly, many researchers had suggested that precipitate strengthening effect was diminished (variation in hardness and strength) owing to FSP (Abu-warda et al. 2021; El-Rayes and El-Danaf 2012; Sharma et al. 2015).

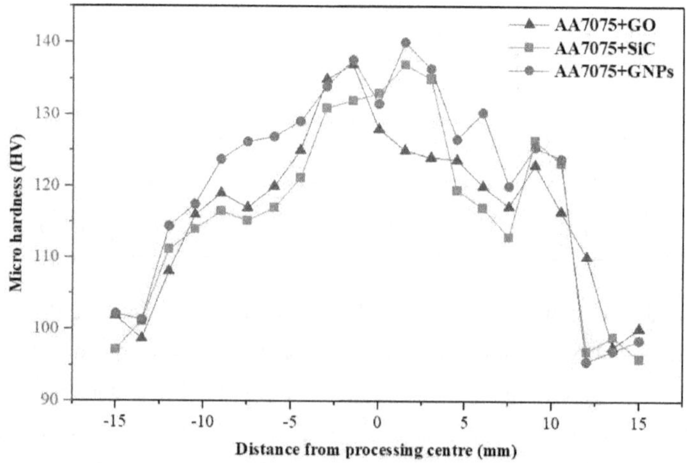

Fig.4.12 Variation in microhardness caused by FSP effect when the tool traverse speed at 25 mm/min

From the Table 4.8 shown, the mechanical properties are strongly depends on the reinforcements, process and tool variables. It observed that TRS of more than 1000 rpm achieved better mechanical properties, irrespective of TTS, improves the material mixing and bonding. The tool with (D/d) ratio of 3 yields and plain cylindrical tool pin profile resulted better UTS for both existing and present work. The tool pin profiles contribute to better stress distribution and enhanced strength. Present work comparatively shows better mechanical properties, especially AA7075+SiC combination. It shows exceptionally well with all existing work

Table 4.8 Comparison of mechanical properties of AA7075 surface co

Ref	Reinforcement	Process variables		Tool variables		
		TRS	TTS	D	d	Tool p profil
(Anandan and Manikandan 2023)	Unreinforced	1000	25, 45, 65 and 85	-	-	TC
(Kumar and Kumar 2023)	SiC	730, 960 and 1190	43, 54 and 65	25	8	PC
(Rana et al. 2016)	B_4C	545	50, 78 and 120	18	3-5	TC
(Jamali and Mirsalehi 2022b)	ZrO_2	1000 and 1200	40	18	6	PC
(Butola n.d.)	B_4C	1000	40	19.95	5	SQ
(Tyagi et al. 2020)	SiC+Aloevera	600 and 900	-	19.95	3.5	SQ
(Khan et al. 2025)	Al_2SiO_5 & MWCNTs	750, 900 and 1050	20,40 and 60	-	-	-
Present work*	SiC	1500	25	18	6	PC
	GNPs	1500	25	18	6	PC
	GO	1500	25	18	6	PC

**Best combinations, D – Shoulder diameter (mm), d – pin diameter (mm), PC – Plain cylindrical, Square.*

4.6 INFLUENCE OF TTS ON METALLURGICAL AND MECHANICAL BEHAVIOUR OF AA7075 SURFACE NANOCOMPOSITES

This topic discuss the influence of TTS on metallurgical and mechanical behaviour of AA7075 surface nanocomposites. The TTS was chosen as 20 mm/min, 25 mm/min and 30 mm/min. UTS, HV, CS and Ɛ were chosen as evaluation criteria. Sections 4.6.1 to 4.6.3 discuss the influence of TTS on metallurgical and mechanical behaviour of AA7075+GO, AA7075+SiC and AA7075+GNPs surface nanocomposites, respectively.

4.6.1 Influence of TTS on metallurgical and mechanical behaviour of AA7075+GO surface nanocomposites

Figure 4.13 depicts the influence of TTS on surface integrity of AA7075+GO surface nanocomposites. As infer from Fig.4.13, TTS affects the metallurgical and mechanical performance. Among TTS, a 25 mm/min traverse speed results in a greater lattice strain than the other speeds. The maximum temperature was obtained at this condition was 324.7 °C, facilitated enhance matrix softening, improved particle dispersion, and strong interfacial bonding between GO and the AA7075. Unlike moderate TTS, low TTS resulted insufficient thermal energy, limiting plastic flow, while higher TTS led to reduced material interaction time, causing non-uniform microstructure and reduced surface integrity performance. The low to medium tool traverse speeds raise the amount of frictional heat produced, which promotes severe plastic deformation (Sathiskumar et al. 2013b). This allows large Ɛ followed by grain size refinement that results better mechanical properties. At 25 mm/min, the compressive residual stress was observed with a mean value of 69.67 MPa, which was benefit for fatigue performance, as it resist crack initiation and propagation. It is likely induced by thermal contraction mismatch between the AA7075 and the GO during cooling.

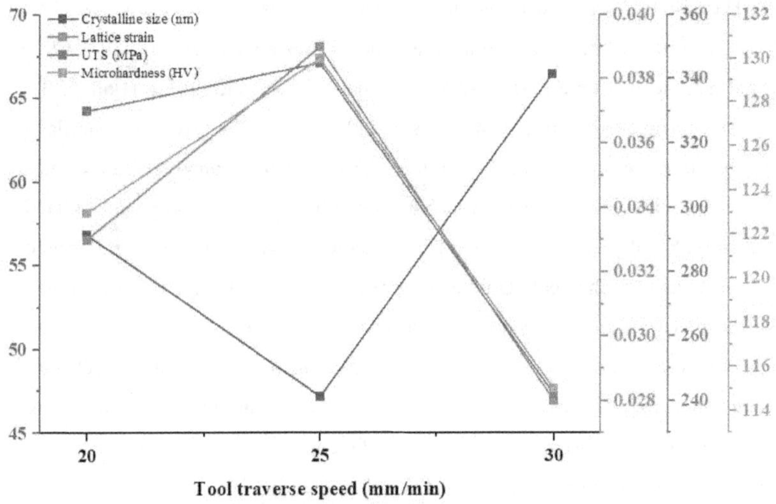

Fig.4.13 TTS vs. CS, Ɛ, UTS and HV for AA7075+GO surface nanocomposite

The TTS of 25 mm/min induced 0.039 and 47.196 nm of Ɛ and CS, correspondingly. A micro hardness of 130 HV and an UTS of 345 MPa are correlated with a refined crystalline size of 47.196 nm. UTS and micro hardness are comparable to those attained at 25 mm/min when the tool traverses at a speed of 20 mm/min. Further, a TTS of 30 mm/min causes a sharp reduction in HV and UTS due to enhanced CS. Unlike other TTS, 25 mm/min of TTS provides an optimal balance of thermal input, residual stress, strength, hardness and microstructural quality, making it is preferable for high-performance, fatigue resistant structural components, include airframe structures.

4.6.2 Influence of TTS on metallurgical and mechanical behaviour of AA7075+SiC surface nanocomposite

Figure 4.14 shows a similar pattern like previous Fig.4.13. As previously stated, mechanical characteristics are inversely correlated with CS. The initial CS is 56.723 nm at the lowest TTS of 20 mm/min. The CS drops to 47.162 nm at 25 mm/min with a difference of 5 mm/min, signifying a 16.85 % improvement. Improved UTS and microhardness of 358 MPa and 139 HV, respectively, were attained at 25 mm/min of TTS. The corresponding maximum temperature was 331.6 °C, promoting effective material softening and uniform distribution of SiC particles within AA7075.

The residual stress measured at moderate TTS was compressive (−78 MPa), likely induced by the changes in cooling and thermal mismatch between the AA7075 and rigid SiC nanoparticles, which constrain matrix shrinkage during cooling. The compressive residual stress is advantageous for enhancing surface properties similar to GO combination. Compared with 30 mm/min, the surface nanocomposite fabricated at 25 mm/min provides improvement in UTS and microhardness of 42.56 % and 13.04 %, consequently. Similar to GO+25, SiC+25 also shows superior properties in terms of metallurgical and mechanical. Unlike moderate TTS, both lower and higher TTS resulted insufficient stirring, leading to coarser grains and reduced metallurgical and mechanical performance. The synergy between refined CS, increase mechanical properties, and compressive residual stress confirms that 25 mm/min of TTS is the most effective TTS for processing AA7075+SiC surface nanocomposites, specifically for demanding structural applications, like GO combination.

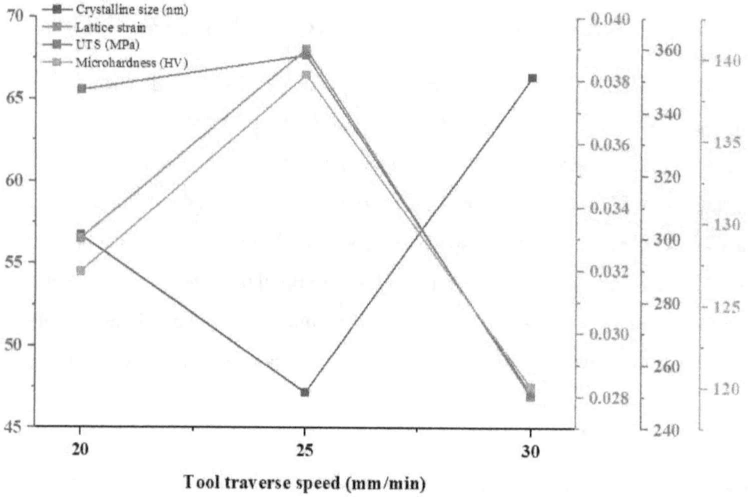

Fig.4.14 TTS vs. CS, ε, UTS and HV for AA7075+SiC surface nanocomposite

4.6.3 Influence of TTS on metallurgical and mechanical behaviour of AA7075+GNPs surface nanocomposite

Figure 4.15 exhibits a similar pattern with reference to Fig.4.14 and Fig.4.13. When processing between 20 and 25 mm/min of TTS, the CS, UTS and HV yielded good results; when processed at a rate of 30 mm/min of TTS, a decrease in properties is observed. A medium traverse speed of 25 mm/min improves the micro hardness,

UTS, and CS. A TTS of 25 mm/min enhances HV and UTS by 6.4% and 3.5%, respectively, over 20 mm/min. In contrast with 30 mm/min of TTS, 25 mm/min significantly increases micro hardness and UTS, with improvements of 13.68 % and 42.85 %, respectively. Similar to AA7075+GO and AA7075+SiC, the AA7075+GNPs surface nanocomposite exhibits better surface integrity performance at a TTS of 25 mm/min. The maximum processing temperature was 328.9 °C, which provided adequate thermal input to promote uniform GNP distribution and strong interfacial bonding. The deviations from this moderate TTS led to reduced surface integrity performance, at 20 mm/min of TTS, limited heat input yielded incomplete recrystallization, while at 30 mm/min of TTS, faster traverse compromised stirring effectiveness and uniformity.

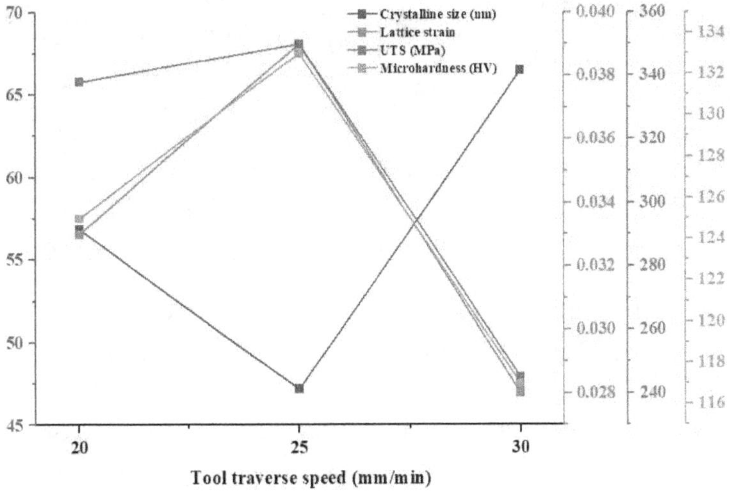

Fig.4.15 TTS vs. CS, Ɛ, UTS and HV for AA7075+GNPs surface nanocomposite

The mean residual stress measured at moderate TTS was compressive, with a magnitude of 73.67 MPa. Compared to SiC, GNPs shows better thermal compatibility with AA7075, contributing to minimal residual stress while maintaining higher surface integrity. Thus, the 25 mm/min TTS offers the optimal combination of refined microstructure, increased mechanical properties, and favourable compressive residual stress, making it a suitable processing condition for structural and fatigue sensitive applications. In all cases, a tool traverse speed of 30 mm/min results in relatively lower surface integrity performances, this is owing to that the tool hits its maximum traverse speed limit, there is a chance of less frictional heat generation, which leads to inadequate stirring, as a result surface integrity declines (Mouli and Rao 2021). The difference in surface integrity of AA7075 surface nanocomposites processed at unique combinations of TTS are displayed in Table 4.9. The effect of reinforcement content within AA7075 at TTS of 20-25 mm/min range produced minimum considerable variation in surface integrity. Beyond 25 mm/min of TTS, the surface integrity shows large variation. A TTS of 20 to 25 mm/min range shows in a variation in R_a (μm) of −0.05, −0.12, and −0.24 for AA7075 reinforced with GO, SiC, and GNPs, respectively. On the other hand, during the 25 to 30 mm/min tool traverse speed range, this disparity is comparatively greater. More specifically, for GO, SiC, and GNPs, the difference in R_a is +0.15, +0.22, and +0.10, respectively.

Table 4.9. Difference in surface integrity at unique combination of TTS

Reinforcement	Tool traverse speed (mm/min)	R_a (μm)	UTS (MPa)	Micro-hardness (HV)	Crystal size (nm)
GO	20-25	−0.05	+15	−07	−09.652
	25-30	+0.15	−103	+15	+19.246
SiC	20-25	−0.12	+11	−12	−09.561
	25-30	+0.22	−107	+19	+19.197
GNPs	20-25	−0.24	+12	−08	−09.670
	25-30	+0.10	−105	+16	+19.283

4.7 EFFCET OF MODERATE TTS ON SURFACE INTEGRITY

Surface integrity plots of AA7075 surface nanocomposites processed at TTS of 25 mm/min are shown in Fig.4.16 (a–d). For fabricating AA7075+SiC surface nanocomposite, the TTS of 25 mm/min produces R_a, UTS, Micro-hardness and crystalline size of 1.53 μm, 358 MPa, 139 HV and 47.162 nm, respectively. This implies that compared to the other nanocomposites, the fabricated AA7075+SiC surface nanocomposite performs better. The surface integrity characteristics of AA7075+GO, AA7075+SiC and AA7075+GNPs showed strong dependence on thermal heat input and residual stress, notably at a moderate TTS. As shown in Fig.4.16 (a-d), the combined properties such as microstructural, mechanical, and surface finish varied with the nature of reinforcement, which correlates with the measured temperature during processing and compressive residual stress developed. In contrast with GO and GNPs, SiC reinforced AA7075 exhibited the highest UTS (358 MPa) and micro hardness (139 HV), along with the lowest R_a (1.53 μm). It records the highest processing temperature and highest compressive residual stress (−78 MPa), those facilitates better material flow and dynamic recrystallization. The elevated temperatures during processing have influenced changes in microstructure and contributed to the residual stress development. Overall, the analysis suggests that processing temperature acts as a governing factor for evolution of microstructure, while compressive residual stress is largely impacted by the thermal expansion mismatch and dynamics of cooling between reinforcement and matrix system. The AA7075+SiC surface nanocomposite emerges as the most favourable in terms of surface integrity, due to its maximum processing temperature and enhanced compressive residual stress. These findings reveal the critical interplay between thermal conditions and stress development during FSP, reinforcing the need to tailor process parameters and reinforcement nature to achieve desired material properties

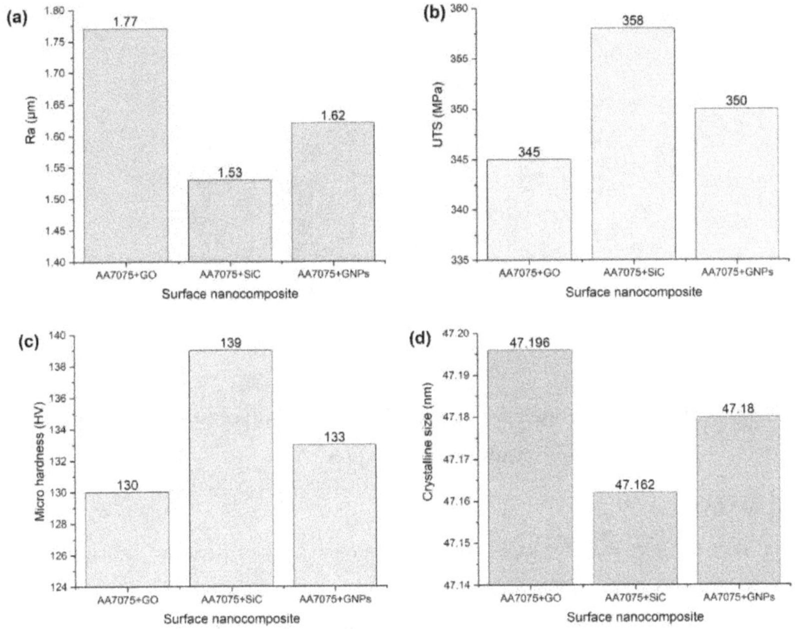

Fig.4.16 Surface integrity plots for fabricated surface nanocomposites at 25 mm/min of tool traverse speed (a) Surface roughness (b) Ultimate tensile strength (c)Micro hardness (d) Crystallite size

4.8 COMPARISON OF MECHANICAL PROPERTIES OF AA7075 SURFACE NANOCOMPOSITE WITH SINTERING TECHNIQUES

The difference in tensile and micro hardness parameters obtained from several techniques, such as spark plasma sintering (SPS), microwave sintering (MWS), conventional sintering (CS), and the current study (FSP), is shown in Fig.4.17 (Manohar et al. 2021). In comparison to previous methods, the present findings shows improved microhardness (139 HV) and tensile strength (358 MPa) under a tool traverse speed of 25 mm/min. With reference to conventional sintering (CS), the present work highlights superior improvement in microhardness and tensile strength of 63.52% and 49.16%, respectively.

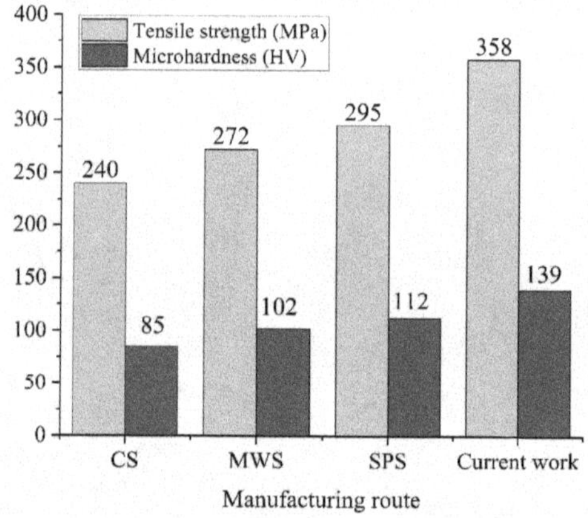

Fig.4.17 Mechanical properties of AA7075 surface nanocomposites versus sintering techniques

4.9 SUMMARY

The influence of tool traverse speed on the surface integrity of friction stir processed AA7075 reinforced with nano graphene oxide, silicon carbide and graphene nanoplatelets were assessed. The following summary have been drawn.

- Tool traverse speed significantly influenced heat generation, with moderate tool traverse speed enhancing heat accumulation, improving material softening and process stability. XRD confirmed the presence of α-Al phase, while FESEM-EDS verified reinforcement incorporation within the matrix.
- SiC reinforcement exhibited highest compressive residual stress, demonstrating effective load transfer and structural strengthening, followed by GNPs and GO.

CHAPTER 5

CORROSION BEHAVIOR OF AA7075 SURFACE NANOCOMPOSITES

5.1 INTRODUCTION

This chapter investigates the corrosion behavior of AA7075 surface nanocomposites under 3.5 wt.% sodium chloride (NaCl) solution. A potentio dynamic polarization test (PDPT) was employed to evaluate the corrosion behaviour of AA7075 surface nanocomposites. The corrosion behaviour of AA7075 surface nanocomposites was structured around two main objectives: first, to study the influence of Tool Traverse Speed (TTS) on Corrosion Rate (CR), and second, to study the impact of specific nano reinforcement combinations (AA7075+GO, AA7075+SiC, and AA7075+GNPs) on CR. The CR was examined statistically using descriptive statistics (mean, standard deviation and variance), normality tests, non-parametric tests, and analysis of variance (ANOVA). Based on statistical analysis, the corroded surfaces were examined using surface morphology and surface topography techniques. Application and comparison of present findings with other works are detailed in the last.

5.2 CORROSION BEHAVIOR ANALYSIS WITH UNIQUE COMBINATION OF TOOL TRAVERSE SPEED USING CYLINDRICAL PIN PROFILE

Fabrication of AA7075 surface nanocomposites were done with unique reinforcements, fixed tool rotational speed (TRS), variable TTS and cylindrical tool pin (TP) profile. Reinforcements were chosen as Graphene Oxide (GO), Graphene nanoplatelets (GNPs) and Silicon carbide (SiC). TRS and TTS were selected as 1500 rpm and 20-30 mm/min, respectively. To analyze the influence of TTS on CR of AA7075 surface nanocomposites, PDPT was done under 3.5 wt. % NaCl. A cathodic potential of −200 mV, an anodic potential of 400 mV, and a scanning speed of 2 mV/s were used as electro chemical operating parameters. The CR was calculated based on Tafel extrapolation method. The CR was further examined statistically using descriptive statistics (mean, standard deviation and variance), normality tests, non-parametric tests, and analysis of variance (ANOVA). Based on statistical analysis, the corroded surfaces were examined using optical microscopy (OM), FESEM-EDS and 3D surface topographic techniques.

5.2.1 Statistical analysis of the CR of AA7075 surface nanocomposites

The CR of different combinations were examined to understand their behavior under same conditions. The readings (R), displayed in Table 5.1, reveals the descriptive statistical data regarding mean, median, SD, and variance (Var).

Table 5.1 Descriptive statistical analysis of the CR of AA7075 surface nanocomposite

Combination	CR (mm/year)			Mean	Median	SD	Variance
	R_1	R_2	R_3				
GO+20	0.0225	0.0288	0.0380	0.0298	0.0288	0.0078	0.0000608
GO+25	0.0610	0.0119	0.0272	0.0334	0.0272	0.0251	0.0006312
GO+30	0.0415	0.0270	0.0106	0.0264	0.0270	0.0155	0.0002390
SiC+20	0.0144	0.0231	0.0490	0.0288	0.0231	0.0180	0.0003239
SiC+25	0.0663	0.0252	0.0661	0.0525	0.0661	0.0237	0.0005603
SiC+30	0.0110	0.0260	0.0252	0.0207	0.0252	0.0084	0.0000712
GNPs+20	0.0689	0.0120	0.0134	0.0314	0.0134	0.0325	0.0010533
GNPs+25	0.0150	0.0207	0.0690	0.034 9	0.0207	0.0297	0.0008802
GNPs+30	0.0120	0.0359	0.0250	0.0243	0.0250	0.0120	0.0001432

For GO combinations, the CR reduces at a TTS of 30 mm/min, suggesting enhanced corrosion resistance at higher speeds. Among the corroded samples, GO+30 shows the lowest CR, whereas GO+25 shows the greatest variability. As per SiC reinforced combination, SiC+30 exhibits the lowest CR with minimal variation, making it most stable surface nanocomposite. Conversely, SiC+25 combination identifies the highest CR. Similarly, for GNPs reinforced combination, a TTS of 30 mm/min achieves the lowest CR, while GNPs+25 combination shows the highest CR with significant variability. The GO+20, SiC+20 and GNPs+20 combinations shows moderate corrosion rates. Across all combinations, a higher TTS of 30 mm/min generally improves corrosion resistance in surface nanocomposites. Table 5.2 displays the mean CR in mm/year across RT and TTS. Among RT, GO showed a mean CR of 0.0298 mm/year unlike SiC. Under TTS combinations, 30 mm/min performed well with a mean CR of 0.0238 mm/year, unlike 25 mm/min (0.0403 mm/year). Among all combinations, GO+30, SiC+25 and GNPs+20 shows the mean CR of 0.026 mm/year, 0.052 mm/year and 0.031 mm/year, respectively. The Q-Q plots for different combinations are shown in Fig.5.1 (a-i). It shows the alignment of ordered values with theoretical quantiles, indicating normal data distribution for most combinations, except SiC+25 (Fig.5.1 e). To validate Q-Q plots, the Shapiro-Wilk test was conducted.

Table 5.2 Mean CR of AA7075 surface nanocomposites across RT and TTS

RT	20 mm/min			25 mm/min			30 mm/min			Mean
GO	0.022	0.028	0.038	0.061	0.011	0.027	0.041	0.027	0.010	0.029
SiC	0.014	0.023	0.049	0.066	0.025	0.066	0.011	0.026	0.025	0.034
GNPs	0.068	0.012	0.013	0.015	0.020	0.069	0.012	0.035	0.025	0.030
Mean	0.0300			0.0403			0.0238			

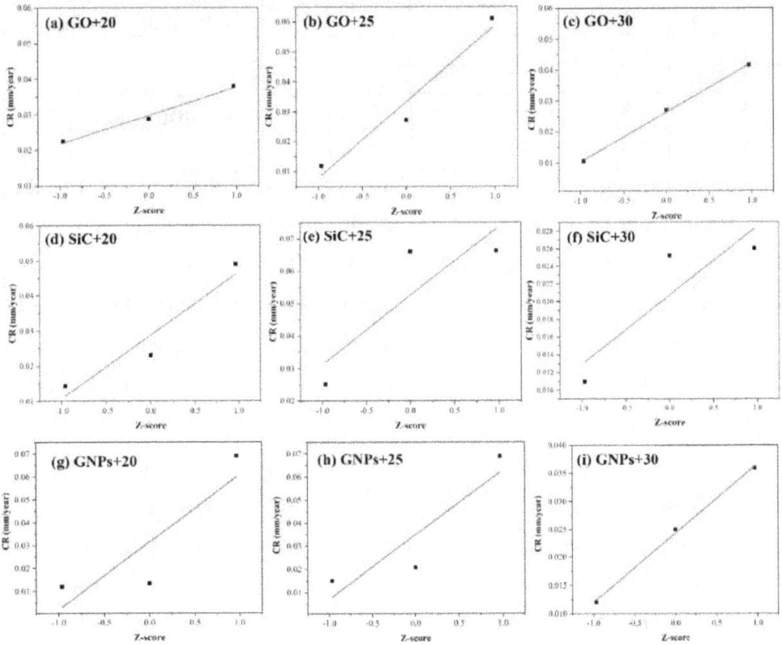

Fig.5.1 (a-i) Quantitle Quantile (Q-Q) plots of the CR of AA7075 surface nanocomposites

All experimental combinations, with the exception of SiC+25 combination ($W = 0.753$, $P < 0.05$), were found to be normally distributed by the Shapiro-Wilk test, suggesting a possible breach of normality. This result agrees with Q-Q plots. Table 5.3 shows the Shapiro-Wilk test results on CR. There was no discernible variation in variances between the combinations, according to Levene's test for homogeneity of variances ($L = 0.255$, $p > 0.05$). As a result, the ANOVA's assumption of equal variances was maintained. Table 5.4 displays the Levene's test results on CR.

Table 5.3 Shapiro-Wilk test of the CR of AA7075 surface nanocomposites

Combination	W	Critical observations from statistical table		
		Critical values	Confidence levels	Normality status
GO+20	0.988	0.959 to 0.998	$P_{0.5}$ to $P_{0.9}$	Data follows normal distribution
GO+25	0.955	0.959 to 0.998	$P_{0.5}$ to $P_{0.9}$	
GO+30	0.998	0.959 to 0.998	$P_{0.5}$ to $P_{0.9}$	
SiC+20	0.924	0.789 to 0.959	$P_{0.1}$ to $P_{0.5}$	
SiC+25	0.753	0.753	$P_{0.01}$	Data doesn't follow normal distribution
SiC+30	0.790	0.789 to 0.959	$P_{0.1}$ to $P_{0.5}$	Data follows normal distribution
GNPs+20	0.768	0.767 to 0.789	$P_{0.05}$ to $P_{0.1}$	
GNPs+25	0.828	0.789 to 0.959	$P_{0.1}$ to $P_{0.5}$	
GNPs+30	0.997	0.959 to 0.998	$P_{0.5}$ to $P_{0.9}$	

The distribution of CR values for all combinations are illustrated in Fig.5.2. Narrow Interquartile (IQR) for GO+20 and SiC+30 shows consistent corrosion performance, unlike the wider IQR for SiC+25. SiC+25 combination shows greater variability. When the combinations are compared, the SiC combination often exhibits greater variability than the GO and GNPs combinations. This demonstrates how reinforcement and TTS affect corrosion resistance. GNPs combinations shows moderate to high variability, with GNPs+30 shows the most stable performance. RT and TTS are significantly affect corrosion resistance, with GO based combinations demonstrating superior stability. The Kruskal-Wallis test ($H = 3.630 < \chi^2_{critcal} = 15.507$, $df = 8$) indicated no significant differences in medians among combinations, aligning with the observed visual overlaps in box plots. This result aligns with the conclusion of Levene's test and it is further supported by the box plots' visual overlaps. Table 5.5 displays Non-parametric test details.

Table 5.4 Levene's test of the CR of AA7075 surface nanocomposites

No. of combinations (k)	Total no. of observations (N)	Degrees of freedom (df)		Calculated test statistics, (L)	Critical value from F-table	Test status
		df_1=N-k	df_2=k-1			
9	27	18	8	0.255	2.51	Fail to reject H_0

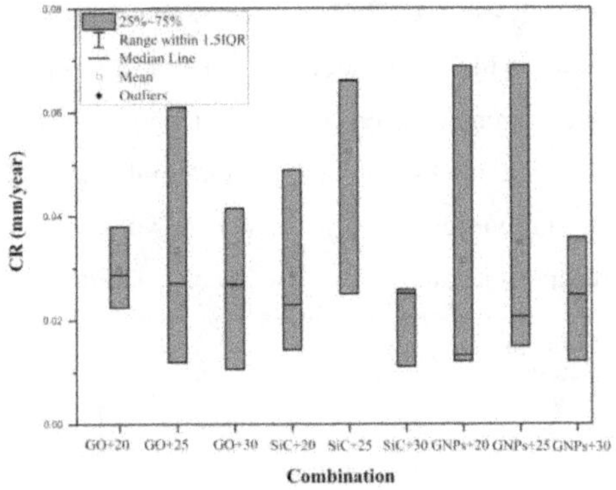

Fig.5.2 Box plot visualization of the CR of AA7075 surface nanocomposite

Table 5.5 Kruskal's Wallis test of the CR of AA7075 surface nanocomposite

No. of combinations (k)	**Degrees of freedom (df) = k-1**	**Calculated test statistics, (H)**	**Critical value from chi-square table**	**Test status**
9	8	3.630	15.507	Fail to reject H_0

After validating the results obtained from normality and homogeneity of variance, ANOVA is being used to compare the combination means. According to ANOVA reports ($F < F_{critcal}$), CR was not substantially impacted by TTS, RT, or their interaction. The ANOVA for CR is displayed in Table 5.6. The findings indicate that the observed differences among combinations are likely due to statistical noise (random fluctuations) rather than true variations in material characteristics. In addition, it may indicate that other factors, such as environment conditions or microstructural characteristics could be more dominant in determining the corrosion rate.

Table 5.6 ANOVA outcomes of the CR of AA7075 surface nanocomposite

Source of variation	**Sum square (SS)**	**df**	**Mean square (MS)**	**Calculated, F**	**Critical value from F-Table**	**Test status**
TTS	0.0012	2	0.00062	1.291	3.55	None of the factors have significant effect on CR
RT	0.0001	2	0.00005	0.104		
TTS*RT	0.0006	4	0.00016	0.333	2.93	
Error	0.0086	18	0.00048			
Total	0.0106	26				

5.2.2 Corroded surface analysis of AA7075 surface nanocomposites

The Tafel plot shown in Fig.5.3 illustrates the relationship between the potential and the logarithm of the current density. The CR of a surface nanocomposite is significantly influenced by both the E_{corr} and the i_{corr} (Karmuhilan and Somasundaram 2023). The CR is directly proportional to i_{corr}. Higher the value of i_{corr} represents the faster electrochemical reactions leading to increase in corrosion rate.

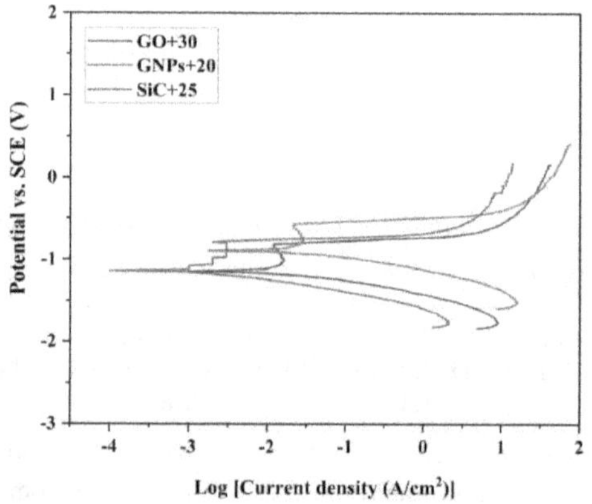

Fig.5.3 Tafel plots of AA7075 surface nanocomposites

GO+30 exhibits lower current densities in both the anodic and cathodic areas than the others, which may indicate increased corrosion resistance. GNPs+20 shows a higher corrosion current density (i_{corr}) with respect to GO+30, suggesting a comparatively higher corrosion rate. In contrast with other combinations, the highest i_{corr} of SiC+25 indicating least corrosion resistance. The E_{corr}, i_{corr} and CR of fabricated surface nanocomposites are displayed in Table 5.7. SiC+25 exhibited highest i_{corr} leading to a highest CR, unlike other surfaces. A more negative corrosion potential indicates greater susceptibility to corrosion initiation. SiC+25 has shown higher susceptibility to corrosion initiation (-1.1675 ± 0.082 V). Grain boundaries (GBs), which serve as active corrosion sites, are denser when there are smaller grains present (Barbucci et al. 1998; Osório et al. 2005; Palumbo et al. 1990, 1991b). GBs have a larger surface energy and reactivity, this typically results in higher corrosion rates. Larger grains improved corrosion resistance and decrease reactivity by reducing the GB area. GO+30 has the largest CS and the lowest CR, indicating better corrosion resistance. Larger grains reduce the GB area, reducing the material's reactivity and improving corrosion resistance, unlike smallest grain size observed in SiC+25.

Table 5.7 E_{corr}, i_{corr} and CR of AA7075 surface nanocomposites

Combination	E_{corr} (V)	i_{corr} (A/cm^2)	CR (mm/year)
GO+30	−1.0020±0.047	$1.030\times10^{-5}\pm7.20\times10^{-6}$	0.0264±0.0150
SiC+25	−1.1675±0.082	$1.480\times10^{-5}\pm0.06\times10^{-6}$	0.0525±0.0237
GNPs+20	−1.0215±0.017	$1.053\times10^{-5}\pm0.02\times10^{-6}$	0.0314±0.0325

The OM and FESEM image of corroded GO+30 as shown in Fig.5.4 (a-b). As shown in Fig.5.4 (a), pitting corrosion is evidence (Wu et al. 2022). The pits are smaller in size and it is scattered, indicating minimal localized corrosion. As per proven reports, smaller pits led to better corrosion resistance (Trdan and Grum 2012). FESEM image shown in Fig.5.4 (b) indicates corroded pits are shallow and less aggressive, confirming reduced material loss. GO acts as a protective barrier, reducing the base alloy's exposure to the NaCl solution. This aligns with earlier studies indicating that carbonaceous reinforcements reduce CR owing to barrier properties (Tan et al. 2025). Fig.5.5 (a-d) shows the FESEM image of the corroded GO+30, along with elemental distribution analyzed using EDS. The Fig.5.5 (a) shows the surface morphology of corroded specimen with EDS spot locations.

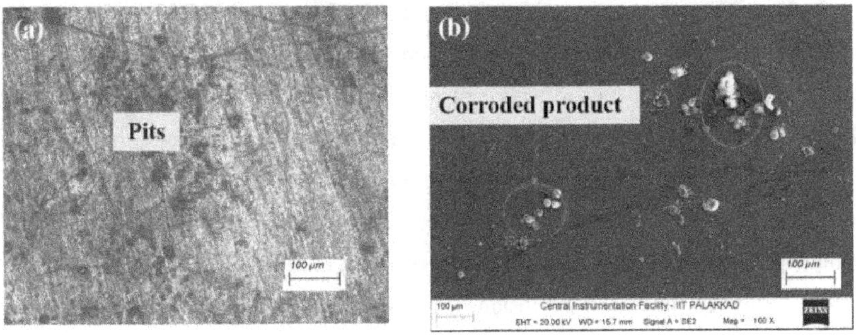

Fig.5.4 Surface morphology of corroded GO+30 surface (a-b) OM and FESEM respectively, shows the formation of pits

The EDS spectrum at the SZ spot shown in Fig.5.5 (b), exhibits the presence of high rich aluminum along with carbon (C) and oxygen (O), confirming the existence of GO reinforcement. The EDS spectrum at the corroded spot shown in Fig.5.5 (c), exhibits significant hike in oxygen by 44.08 wt.%, revealing intense oxidation due to corrosive attack. Al content reduces to 54.38 wt.%, indicating degradation of material and formation of oxide layer. The presence of Magnesium (Mg), Silicon (Si), Copper (Cu) and Zinc (Zn) signifies the localized corrosion induced phase segregation.

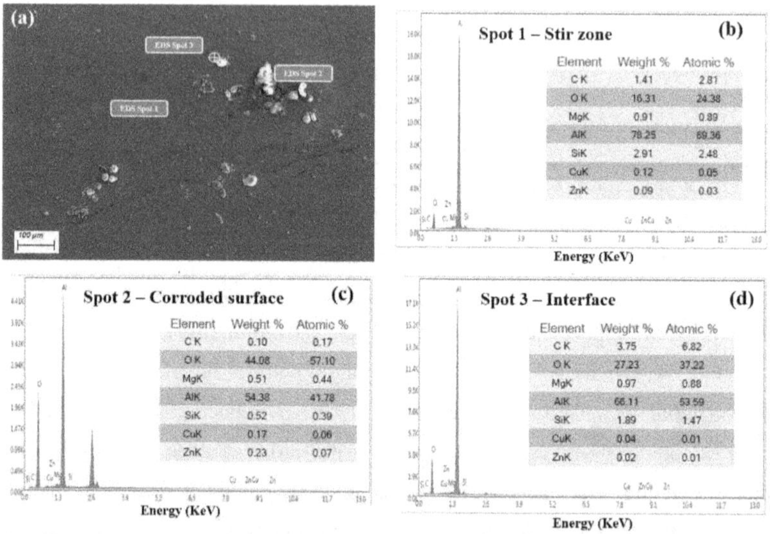

Fig.5.5 FESEM analysis of corroded GO+30 (a) EDS spots on various zones (b) EDS results at SZ (c) EDS results at corroded spot (d) EDS results at interface

The interface region as shown in Fig.5.5 (d), exhibits 27.23 wt.% of O concentration, which is lower than that of corroded surface but higher than the SZ, revealing moderate oxidation. The C content of 3.75 wt.% confirms the presence of GO, potentially enhancing corrosion resistance at interface. The findings are aligned with earlier reports that the aluminum alloys reinforced with graphene reduce CR owing to the formation of protective layer (Badr et al. 2019). With addition, the relative amount of Al by 66.11 wt.% suggests partial retention of the SZ composition. GO+30 surface exhibited increased corrosion resistance, unlike SiC reinforced surface nanocomposites, attributed to the oxygen-containing functional groups such as hydroxyl, carboxyl, and epoxy. These functional groups enhance the distribution of GO particles within the AA7075 matrix. These polar groups promoted electrostatic interactions and hydrogen bonding, ensuring a uniform distribution of GO and preventing the effect of agglomeration.

The interfacial bonding between GO and the AA7075 matrix is also strengthened through these functional groups, which form chemical bonds with the aluminium atoms, increasing the overall stability of the interface. Additionally, hydroxyl, carboxyl, and epoxy play a critical role in the passivation process by promoting the formation of a dense and uniform passive oxide layer on the composite surface, which acts as a barrier protective against corrosion. The functional groups on GO decrease the electrochemical reactivity at the GBs, contributing to a reduced CR and increased long-term performance (Badr et al. 2019; Du et al. 2021; Obaid and Al-Bermany 2023; Tan et al. 2025). The OM and FESEM image of corroded SiC+25 as shown in Fig.5.6 (a-b). As shown in Fig.5.6 (a), pits are larger and deeper across the surface, indicating severe pitting corrosion. FESEM image shown in Fig.5.6 (b) shows well defined corroded pits with notable material loss, indicating localized attack. With reference to surface morphological observations, SiC reinforcement can introduce interfacial imperfections that serve as favorable corrosion sites and does not form a passive layer. As per proven study, micro-galvanic corrosion between aluminum matrix and SiC particles may accelerate corrosion (Dwivedi and Sahu 2018; Loto and Babalola 2017). Fig.5.7 (a-d) shows the FESEM image of the corroded AA7075+SiC surface nanocomposite processed at 25 mm/min of TTS, along with elemental distribution analyzed using EDS.

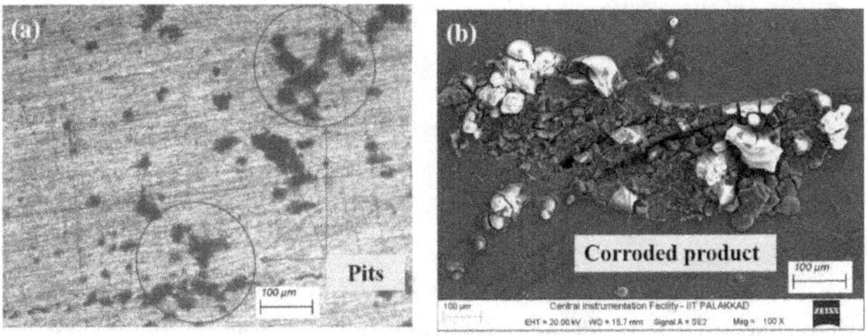

Fig.5.6 Surface morphology of corroded SiC+25 surface (a) OM showed larger pits (b) FESEM reveals well defined corroded pits

Figure 5.7 (a) illustrates the surface morphology of corroded SiC+25 with EDS spot locations. The EDS spectrum at the SZ spot shown in Fig.5.7 (b), indicates a predominant Al content, highlighting the Al-rich matrix of AA7075. The moderate amount of O (18.92 wt.%) contamination reveals partial oxidation. An addition of Si (3.64 wt.%) further confirms the presence of SiC reinforcement. The EDS spectrum at the corroded spot shown in Fig.5.7 (c), exhibits high O (54.66 wt.%), indicating intense oxidation and corrosion induced degradation. A notable decline in Al shows surface material loss owing to corrosion. The presence of Si along with other elements may be due to selective corrosion, leads to phase separation and oxide formation. The interface region as shown in Fig.5.7 (d), exhibits moderate O levels, suggesting partial oxidation but less severe than the corroded spot. The Si wt.% is relatively higher, confirming the existence of SiC reinforcement near the interface. The Al concentration remains high, indicating a relatively intact matrix. SiC reinforcement improved tribological property but had limited corrosion resistance owing to Al-SiC interfacial reactivity (Ikubanni et al. 2022; Pradhan et al. 2016). The refined CS (47.162 ± 0.054 nm) of AA7075+SiC led to increase the proportion of GBs, which can decline the corrosion resistance due to the higher energy and chemical activity of GBs.

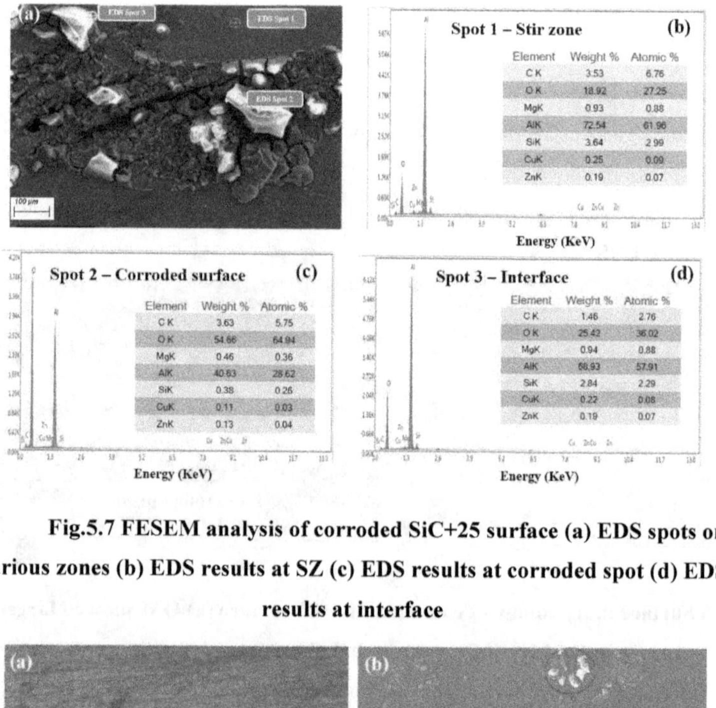

Fig.5.7 FESEM analysis of corroded SiC+25 surface (a) EDS spots on various zones (b) EDS results at SZ (c) EDS results at corroded spot (d) EDS results at interface

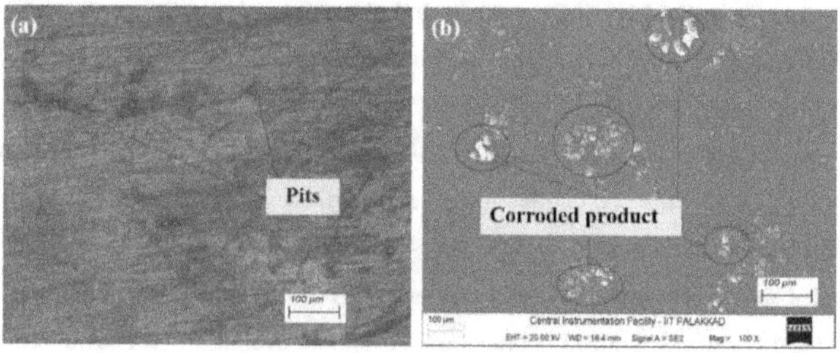

Fig.5.8 Surface morphology of corroded GNPs+20 surface (a) OM shows pits (b) FESEM reveals corroded pits

The OM and FESEM image of corroded GNPs+20 as shown in Fig.5.8 (a-b). As shown in Fig.5.8 (a), pits are visible but less severe than SiC. FESEM image shown in Fig.5.8 (b) reveals that the corroded pits are deeper than those in GO+30 but shallower than those in SiC+25. Due to high conductive nature of GNPs, it promotes localized electrochemical reactions, making GNPs+20 prone to galvanic corrosion. However, the extent of corrosion is minor than SiC+25 owing to partial passivation by GNPs (Achutarao et al. 2022; Xie et al. 2021). Fig.5.9 (a-d) shows the FESEM image of the corroded GNPs+20, along with elemental distribution analyzed using EDS.

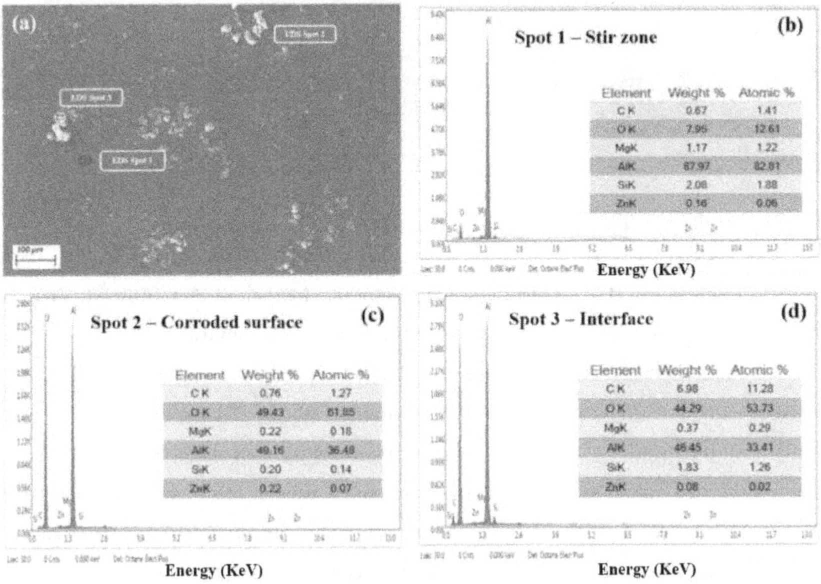

Fig.5.9 FESEM analysis of corroded GNPs+20 surface (a) EDS spots on various zones (b) EDS results at SZ (c) EDS results at corroded spot (d) EDS results at interface

Figure.5.9 (a) illustrates the surface morphology of corroded GNPs+20 with EDS spot locations. The EDS spectrum at the SZ spot shown in Fig.5.9 (b), indicates a predominant Al (87.97 wt. %) content, highlighting the Al-rich matrix of AA7075. The moderate amount of O (7.95 wt.%) contamination reveals initial oxidation. The presence of C (0.67 wt.%), suggesting potential incorporation of GNPs. The EDS spectrum at the corroded spot shown in Fig.5.9 (c), exhibits high O (49.43 wt. %), indicating severe oxidation and corrosion induced degradation.

The O content is slightly lower than SiC reinforcement, indicating GNPs+20 offers better corrosion resistance than SiC+25 but less than GO+30. GNPs can serve as a physical barrier against hostile ions, reducing corrosion rate (Cataldi et al. 2018). The corrosion mechanism under 3.5 wt. % NaCl solution for aluminum surface composites is depicted in Fig.5.10.

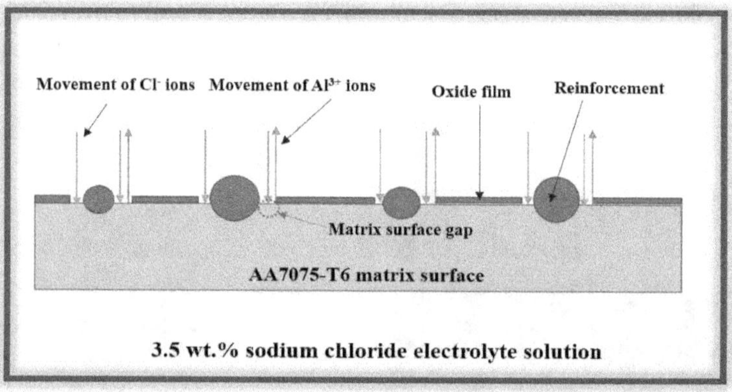

Fig.5.10 Schematic representation of corrosion mechanism of aluminum surface nanocomposite happens in sodium chloride solution

The equations (5.1-5.4) represent an electrochemical workstation's cathodic and anodic reactions (Reboul and Baroux 2011; Sunitha et al. 2019; Zhang et al. 2021). The base metal is exposed to further corrosion owing to the unequal adsorption of chloride ions on the passivation layer, which results in pores. Most of this damage happens in contact between the matrix and the reinforcement. The disintegration of the matrix and reinforcing particles causes pitting, as seen in the corroded samples.

$$2H^{+}+2e \rightarrow H_2 \quad (5.1)$$

$$O_2+2H_2O+4e \rightarrow 4(OH)^{-} \quad (5.2)$$

$$Al \rightarrow Al^{3+}+3e^{-} \quad (5.3)$$

$$Al^{3+}+3Cl^{-} \rightarrow AlCl_3+3e^{-} \quad (5.4)$$

Pore enlargement brought on by the anodic reaction raises the concentration of Al^{3+}, allowing chloride ions to diffuse and start hydrolysis. The hydrolysis reaction is given by equation 5.5, which implies that $Al(OH)_3$ and other corrosion products are deposited on the pores, obstructing cells and exacerbating corrosion (Zhang et al. 2021). According to EDS mapping, elements like as C, O, Mg, Al, Si, Cu and Zn are present in the corrosion products, and same was proved elsewhere (Kandasamy et al. 2020).

$Al^{3+} + 3H_2O \rightarrow Al(OH)_3 + 3H^+$ (5.5)

Figure 5.11 (a-c) shows the surface topography of corroded GO+30. As shown in Fig.5.11 (a), plot is drawn between roughness distance (X-axis) and roughness height (Y-axis). The surface irregularities are visualized by peaks and valleys, average roughness (R_a) of a corroded pit is 136.5 nm (0.136 µm). Figure 5.11 (b) shows 3D topographical view of overall corroded GO+30 surface, with height variations denoted in different colors. The height variations observed between 0 and 63.1 µm. The surface dimensions are denoted by X and Y, visualizing widespread roughness distribution owing to corrosion. A magnified view of a detailed corroded pit is shown in Fig.5.11 (c), reveals the height difference within the pit is upto 51 µm, indicating material degradation.

Figure 5.12 (a-c) shows the surface topography of corroded SiC+25. As shown in Fig.5.12 (a), R_a of a corroded pit is measured at 2.045 µm, which shows higher than that of GO+30 surface. Figure 5.12 (b) shows 3D topographical view of the overall corroded SiC+25 surface, showing height variations observed from 0 to 154 µm, indicating more than twice the height observed in the GO+30 surface. The rough and uneven structure indicating severe localized corrosion attacks. A detailed view of a corroded pit is shown in Fig.5.12 (c) visualize the height difference of upto 0.12 mm, indicating more severe degradation than GO based reinforcement. The depth and surface roughness within the pit further indicate significant material degradation. Fig.5.13 (a-c) shows the surface topography of corroded GNPs+20. As shown in Fig.5.13 (a), R_a of a corroded pit is measured at 1.713 µm, which falls between GO+30 and SiC+25 surfaces.

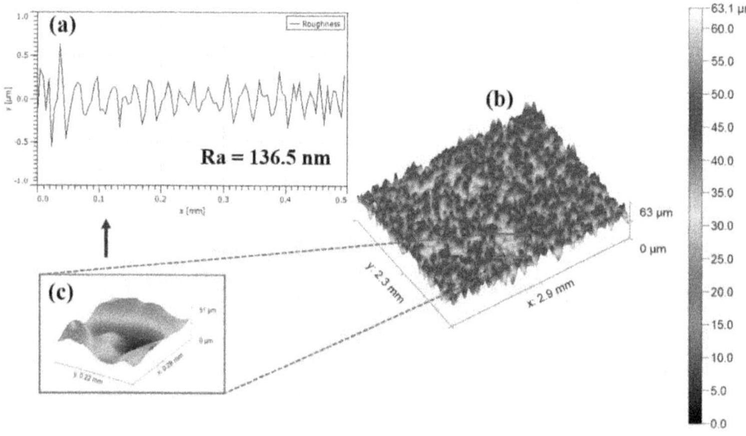

Fig.5.11 Surface topography of corroded GO+30 (a) Surface roughness plot of a pit (b) Overall corroded surface (c) Magnified view of a pit presence in a corroded surface

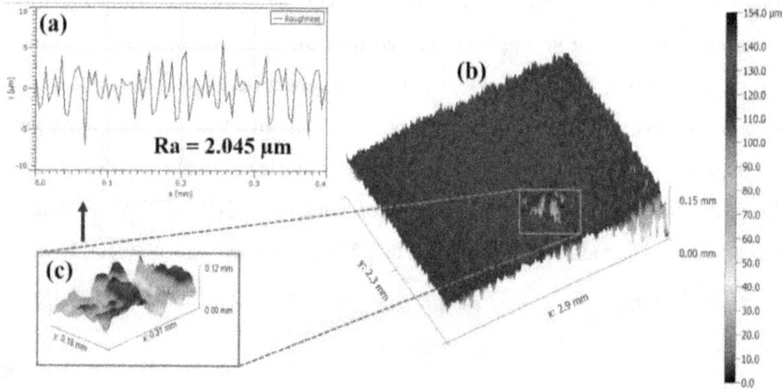

Fig.5.12 Surface topography of corroded SiC+25 (a) Surface roughness plot of a pit (b) Overall corroded surface (c) Magnified view of a pit presence in a corroded surface

Figure 5.13 (b) shows 3D topographical view of overall corroded GNPs+20 surface, with height variations observed between 0 and 136.7 µm. The rough and uneven structure indicating localized corrosion attacks, though with a slightly lower maximum depth compared to the SiC+25 surface. A detailed view of a corroded pit is shown in Fig.5.13 (c) visualize the height difference of upto 0.13 mm, further highlighting the extent of material degradation.

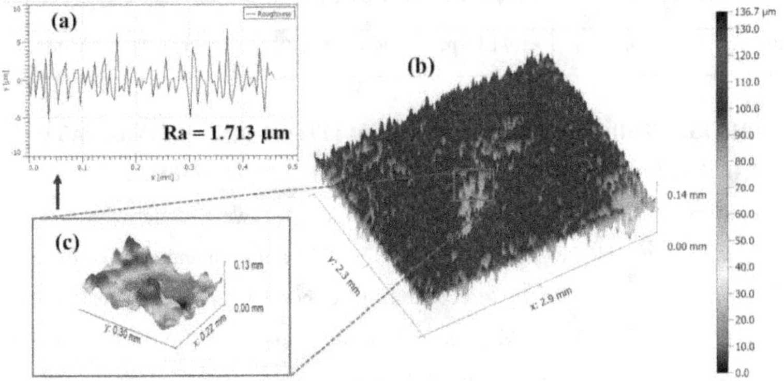

Fig.5.13 Surface topography of corroded GNPs+20 (a) Surface roughness plot of a pit (b) Overall corroded surface (c) Magnified view of a pit presence in a corroded surface

Comparison of surface topography among corroded samples are displayed in Table 5.8. GO+30 produced the least R_a values (both maximum and pitting), shallower pits, a smoother corroded surface morphology and the best corrosion resistance. In contrast, the SiC+25 exhibited the most severe corrosion damage with the highest surface roughness and deepest pits. The corroded surface morphology of GNPs+20 lies between the GO+30 and SiC+25 surfaces. In AA7075 surface nanocomposites, the interplay between reinforcement type, grain size, and lattice strain are significantly influenced by FSP and greatly affects GB formation and stability, with direct implications for the long-term corrosion performance. The CS of the AA7075 matrix is inversely proportional to the ε, as fine CS typically has higher ε owing to DRX during FSP. SiC reinforcement results in a finer CS (47.162 nm), leading to enhanced ε (0.039) and an enhancement in GB density, which promotes localized corrosion due to enhanced chemical reactivity at these boundaries. In contrast, GO reinforcement induces less ε (0.028) and results in a larger CS (66.442 nm), contributes to more stable GBs, increasing passivation and offering better corrosion resistance in long-term use.

Table 5.8 Comparison of corroded surface topography of AA7075 surface nanocomposites

Combination	Average roughness (Ra)		Pit nature	Surface nature	Corrosion resistance
	Maximum	Pitting			
GO+30	63.1 μm	136.5 nm	Shallower	Less rough	Best
SiC+25	154 μm	2.045 μm	Deep and aggressive	Very rough	Least
GNPs+20	136.7 μm	1.713 μm	Moderate	Rough	Moderate

5.3 CORROSION BEHAVIOR ANALYSIS WITH UNIQUE COMBINATION OF REINFORCEMENT TYPE USING THREADED PIN PROFILE

Fabrication of AA7075 surface nanocomposites were done with unique reinforcements, fixed TRS of 1480 rpm and TTS of 30 mm/min and employed using threaded cylindrical TP profile. Reinforcements were chosen as GO, Graphene (Gr) and SiC. XRD, FESEM-EDS, OM were done for fabricated AA7075 surface nanocomposites. Then, to analyze the influence of nano reinforcement on CR of AA7075 surface nanocomposites, PDPT was done under 3.5 wt. % NaCl. A cathodic potential of −200 mV, an anodic potential of 400 mV, and a scanning speed of 2 mV/s were used as electro chemical operating parameters. The CR was calculated based on Tafel extrapolation method. The CR was further examined statistically using descriptive statistics (mean, standard deviation and variance), normality tests, non-parametric tests, and ANOVA. Based on statistical analysis, the corroded surfaces were examined using OM, FESEM-EDS and 3D surface topographic techniques.

5.3.1 Examination of XRD, FESEM-EDS and OM

Figure 5.14 (a-c) shows the XRD patterns of fabricated surface nanocomposites. GO, SiC and Gr combinations are shown in Fig. 5.14 (a), Fig. 5.14 (b) and Fig.5.14 (c), respectively. The prominent peaks associated with α-Al solid solution with h k l indices (1 1 1, 2 0 0 , 2 2 0 , and 3 1 1) are detected. Peaks corresponding to reinforcements are not detected due to lower volume fraction of reinforcements used in this work. The CS and ε were estimated and it is summarized in Table 5.9. The mean CS for SiC, Gr, and GO reinforced surface nanocomposites was 56.24 nm, 45.21 nm, and 52.97 nm, respectively.

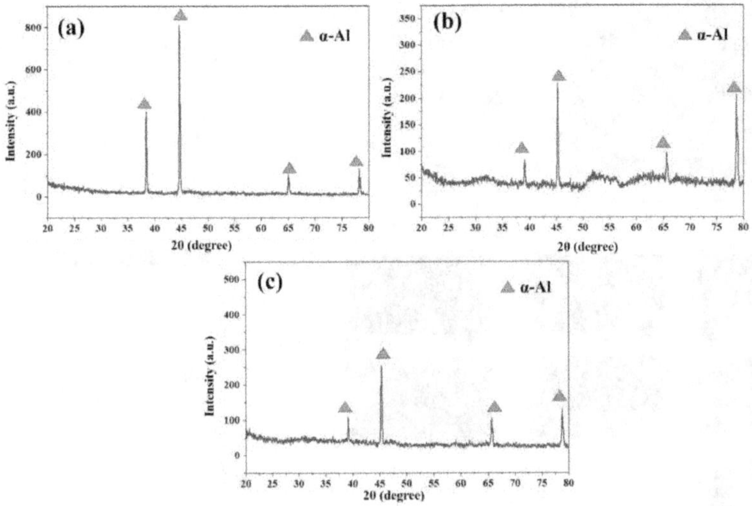

Fig.5.14 XRD examination of AA7075 surface nanocomposites (a) AA7075+GO (b) AA7075+SiC (c) AA7075+Gr

The smallest CS was observed for Gr reinforcement, showing grain refinement, while SiC exhibited the largest CS, indicating enhanced stability. In terms of Ɛ, the Gr reinforced surface nanocomposite showed the highest value, indicating significant lattice distortion. The GO-reinforced surface nanocomposite exhibited a moderate Ɛ, while SiC showed the lowest Ɛ, confirming its role in stabilizing the lattice structure. These findings are aligned with the XRD patterns, where SiC peaks shows sharp and stable, graphite peaks showed intense but distorted, and GO peaks were broadened due to disorder.

Table 5.9 CS and Ɛ of AA7075 surface nanocomposites

Combination	Crystallite size (nm)	Lattice strain
AA7075+GO	52.972 ± 3.465	0.036±0.0020
AA7075+SiC	56.235 ± 2.058	0.033±0.0021
AA7075+Gr	45.212 ± 0.522	0.040±0.0015

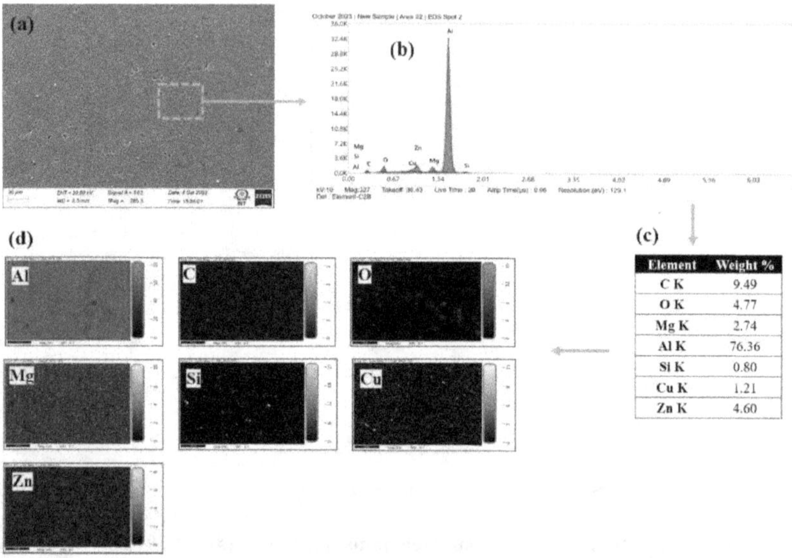

Element	Weight %
C K	9.49
O K	4.77
Mg K	2.74
Al K	76.36
Si K	0.80
Cu K	1.21
Zn K	4.60

Fig.5.15 FESEM analysis of AA7075+GO surface nanocomposite (a) FESEM image with EDS spot (b) EDS results (c) Weight % of elements present (d) Elemental mapping

FESEM-EDS with elemental mapping of AA7075+GO surface nanocomposite is depicted in Fig.5.15 (a-d). The FESEM image of the fabricated surface nanocomposite is shown in Fig.5.15 (a), and its EDS spot analysis is depicted in Fig.5.15 (b). The presence of carbon (C) (9.49 wt.%) and oxygen (O) (4.77 wt.%), along with other elements in the matrix alloy, confirms the presence of GO particles. Fig.5.15 (c) shows the elemental weight % contribution in which C and O are dominant except aluminium. The mapping of Aluminium (Al), C, O, Magnesium (Mg), Silicon (Si), Copper (cu) and Zinc (zn) elements are depicted in Fig.5.15 (d). The FESEM-EDS with elemental mapping of AA7075+SiC surface nanocomposite is shown in Fig.5.16 (a-d). The FESEM image of ceramic reinforcement (SiC) with AA7075 surface is shown in Fig.5.16 (a).The fabricated surface nanocomposite elements are verified using EDS analysis, as shown in Fig.5.16 (b). The combination of Si and C elements contributes 10.73 % by weight. Their existence is much higher than that of other elements except aluminum. Fig.5.16 (c) shows the various elemental contribution by weight %. The elemental mapping of those elements is depicted in Fig.5.16 (d).

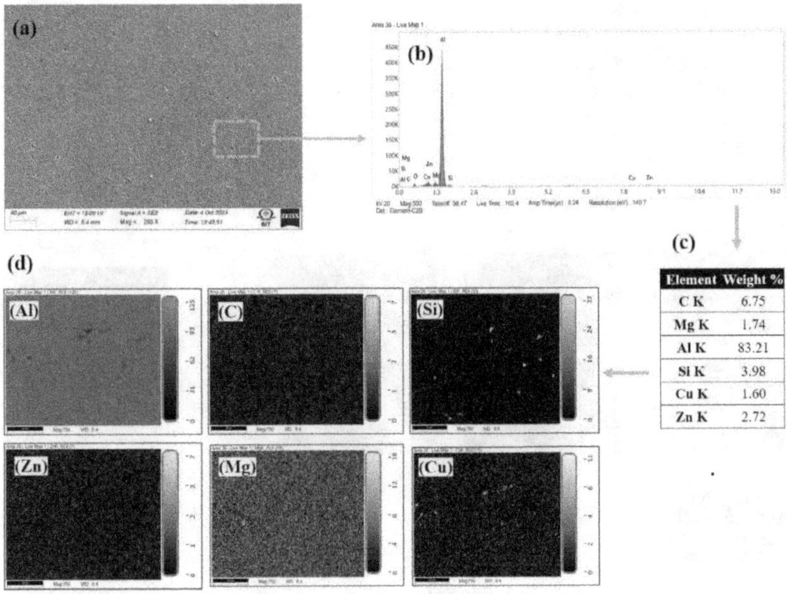

Element	Weight %
C K	6.75
Mg K	1.74
Al K	83.21
Si K	3.98
Cu K	1.60
Zn K	2.72

Fig.5.16 FESEM analysis of AA7075+SiC surface nanocomposite (a) FESEM image with EDS spot (b) EDS results (c) Weight % of elements present (d) Elemental mapping

The FESEM-EDS with elemental mapping of AA7075+Gr surface nanocomposite is depicted in Fig.5.17 (a-d). The FESEM image of AA7075+Gr surface nanocomposite is shown in Fig.5.17 (a). EDS analysis of AA7075+Gr surface nanocomposite is depicted in Fig.5.17 (b-c). As inferred from EDS analysis, the presence of C contributes 38.12 % by weight and O contributes 4.19 % by weight. This confirms the existence of graphene particles. From Fig.5.17 (c), Al, C, O and other elements contribute 52.92 %, 38.12 %, 4.19 %, 42.31 % and 4.77 % by weight, respectively. The mapping of different elements present in the surface nanocomposite is shown in Fig.5.17 (d).

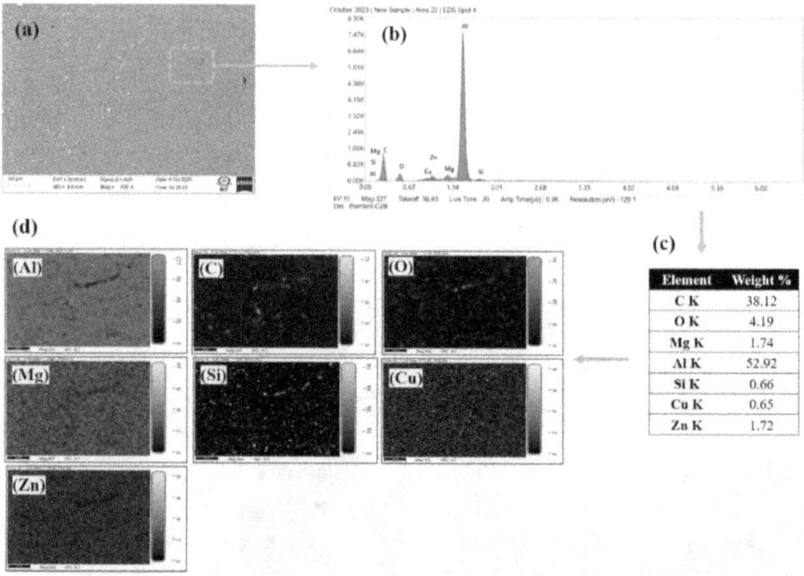

Element	Weight %
C K	38.12
O K	4.19
Mg K	1.74
Al K	52.92
Si K	0.66
Cu K	0.65
Zn K	1.72

Fig.5.17 FESEM analysis of AA7075+Gr surface nanocomposite (a) FESEM image with EDS spot (b) EDS results (c) Weight % of elements present (d) Elemental mapping

The optical microscopic image of AA7075-T6 alloy and AA7075+Gr surface nanocomposite are depicted in Fig.5.18 (a) and Fig.5.18 (b), respectively. The AA7075-T6 shows coarse elongated grains and these grains are refined owing to the effect of severe plastic deformation after FSP. The refinement in grains is mainly ascribed to the addition of graphene particles. Away from Stir zone (SZ), the grains are independently oriented and getting coarser. The Heat affected zone (HAZ) shows more elongated coarse grains than SZ and Thermo mechanically affected zone (TMAZ). These elongated grains deteriorated mechanical properties than refined grains (Mehdi and Mishra 2021). The clear interface is visible for all fabricated surface nanocomposites and it is shown in Fig.5.19 (a-d). The optical microscopy of surface nanocomposite with clear interface is shown in Fig.5.19 (a). AA7075 reinforced with graphene oxide, silicon carbide and graphene show clear interface and it is shown in Fig.5.19 (b), Fig.5.19 (c) and Fig.5.19 (d), respectively. The FESEM images of fabricated surface nanocomposites also shows defect (voids, cavities and cracks) free microstructure, yields better outcomes in terms of surface properties

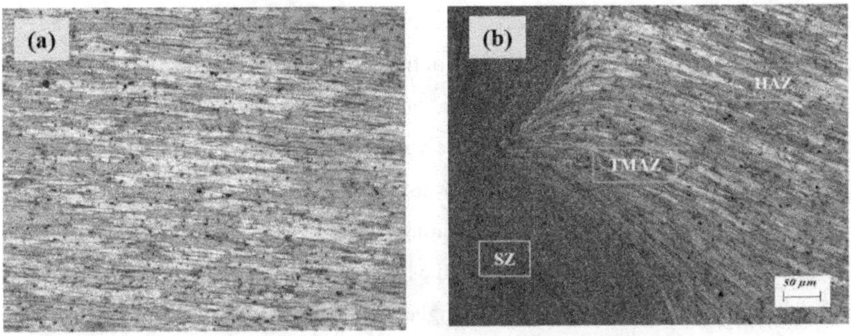

Fig.5.18 Optical micrograph of (a) AA7075-T6 alloy (b) AA7075+Gr surface nanocomposite

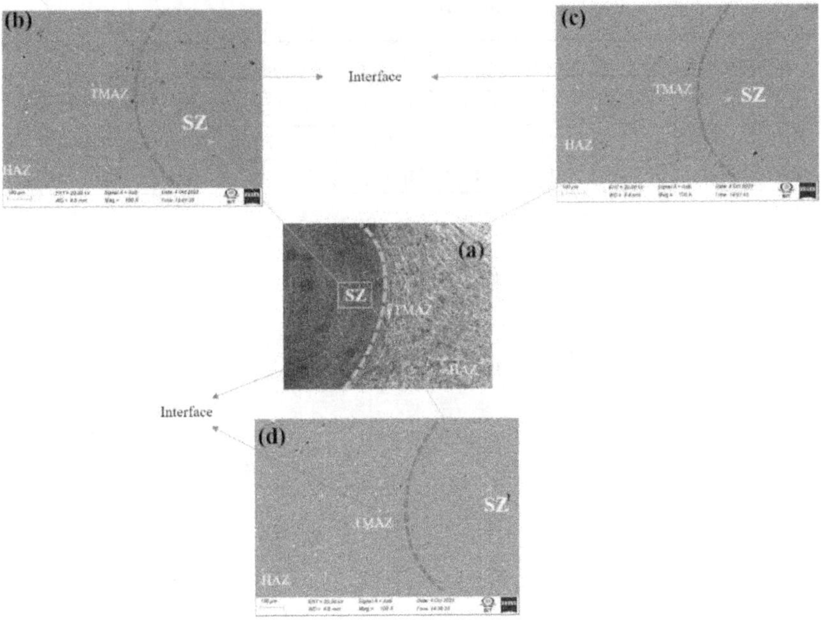

Fig.5.19 Surface morphology of AA7075 surface nanocomposite (a) OM shows AA7075 surface nanocomposite; (b-d) FESEM image of AA7075 reinforced with GO, SiC and Gr, respectively.

5.3.2 Statistical analysis of corrosion rate

Table 5.10 displays the descriptive statistical analysis include mean, median, standard deviation (SD), and variance (Var) of the CR in mm/year of aluminium surface nanocomposites. The AA7075+GO combination showed a mean CR of 0.297 mm/year, with moderate SD of 0.267, suggesting some variability in corrosion performance. The AA7075+SiC combination indicates the lowest mean CR of 0.101 mm/year and the lowest variability with SD of 0.107, highlighting it performs consistently with minimal corrosion. In contrast, the AA7075+Gr had the highest mean CR of 0.918 mm/year and the highest SD of 0.889, indicating both a higher average CR and greater variability in its corrosion performance. This suggests that Gr combination showed less reliable in terms of corrosion resistance.

Table 5.10 Descriptive statistical analysis of the CR of AA7075 surface nanocomposite

Combination	CR (mm/year)			Mean	Median	SD	Var
AA7075+GO	0.297	0.285	0.267	0.2830	0.2850	0.0151	0.0002280
AA7075+SiC	0.101	0.115	0.107	0.1077	0.1070	0.0070	0.0000493
AA7075+Gr	0.918	1.092	0.889	0.9663	0.9180	0.1098	0.0120543

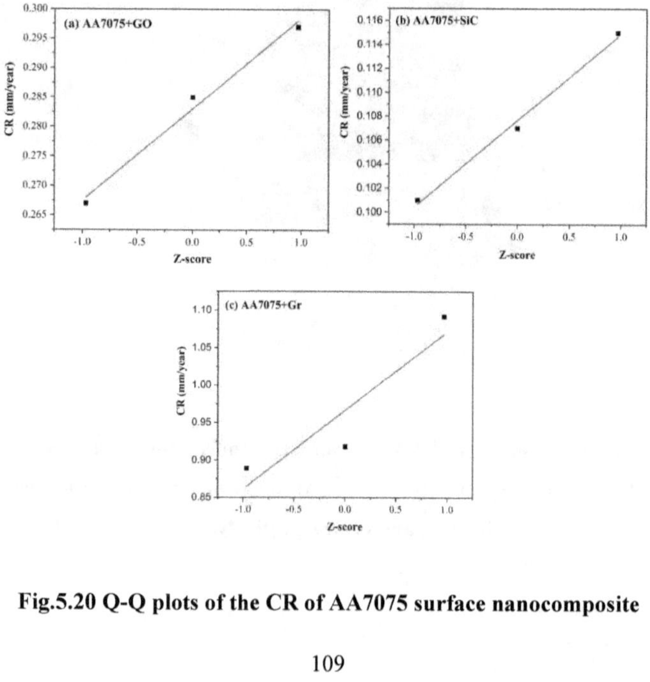

Fig.5.20 Q-Q plots of the CR of AA7075 surface nanocomposite

The Q-Q plots for different combinations are shown in Fig.5.20 (a-c). As shown in Fig.5.20 (a-b), GO and SiC combinations exhibited closely to the reference line, supporting the normality. Gr combination showed slightly deviated from reference line (Fig.5.20 c). To further confirm the normality, Shapiro-Wilk test was conducted. All material combinations, were found to be normally distributed by the Shapiro-Wilk test, suggesting that the data is normally distributed. This result agrees with Q-Q plots. Table 5.11 shows the Shapiro-Wilk test results on CR.

Table 5.11 Shapiro-wilk test of the CR of AA7075 surface nanocomposite

Combination	W	Critical observations from statistical table		
		Critical values	Confidence levels	Normality status
AA7075+GO	0.987	0.959 to 0.998	$P_{0.5}$ to $P_{0.9}$	Data follows normal distribution
AA7075+SiC	0.993	0.959 to 0.998	$P_{0.5}$ to $P_{0.9}$	
AA7075+Gr	0.854	0.789 to 0.959	$P_{0.1}$ to $P_{0.5}$	

There was no discernible variation in variances between the combinations, according to Levene's test for homogeneity of variances ($L = 1.763$, $p > 0.05$). As a result, the ANOVA's assumption of equal variances was maintained. Table 5.12 displays the Levene's test results on CR.

Table 5.12 Levene's test outcomes of the CR of AA7075 surface nanocomposite

No. of combinations (k)	Total no. of observations (N)	Degrees of freedom (df)		Calculated test statistics, (L)	Critical value from F-table	Test status
		df_1=N-k	df_2=k-1			
3	9	6	2	1.763	8.81	Fail to reject H_0

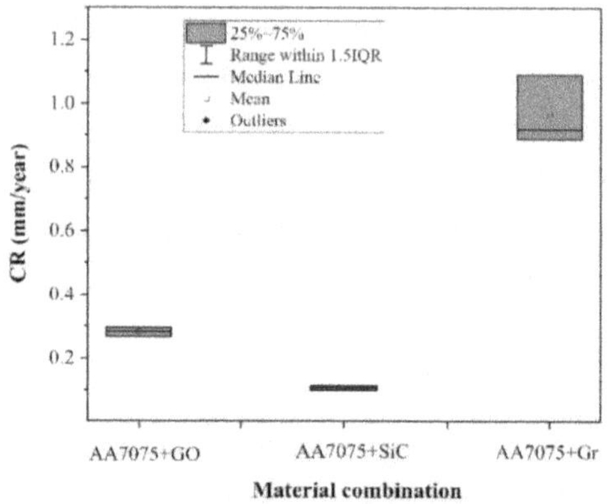

Fig.5.21 Box plots of the CR of AA7075 surface nanocomposites

Box plots of the CR of AA7075 surface nanocomposites are shown in Fig.5.21, reveal nano reinforcement strongly affects the CR. The AA7075 + SiC surface composite showed the best corrosion resistance, with a very low CR and minute variation between samples. The narrow IQR further indicates that SiC reinforcement provides highly consistent and stable corrosion protection. The CR of AA7075+ GO surface nanocomposite lies in between SiC and Gr. The AA7075 + Gr surface nanocomposite shows the poorest corrosion resistance, with a CR close to 1.0 mm/year with greater variation. Table 5.13 displays one-way ANOVA, was conducted to find whether there were statistical difference exists between the means of corrosion rates. The calculated $F_{(2,7)}$ is 172.9, which is very larger than $F_{critcal}$ of 4.74, imply that the CR was highly impacted nano reinforcement addition.

Table 5.13 ANOVA outcomes of the CR of AA7075 surface nanocomposites

Source of variation	SS	df	MS	Calculated, F	F-Table	Test status
Between CR	1.235	2	0.618	172.9	4.74	Significant effect on CR
Within CR	0.025	7	0.004	*		
Total	1.26	9	*	*	*	

5.3.3 Corroded surface analysis of AA7075 surface nanocomposites

The effect of nano particle reinforcement on corrosion mechanisms is intimately related to the changes in microstructure due to their incorporation and distribution within matrix material during FSP. The microstructural results shows that nano particles (GO, SiC and Gr) significantly refined the grain size within the SZ.

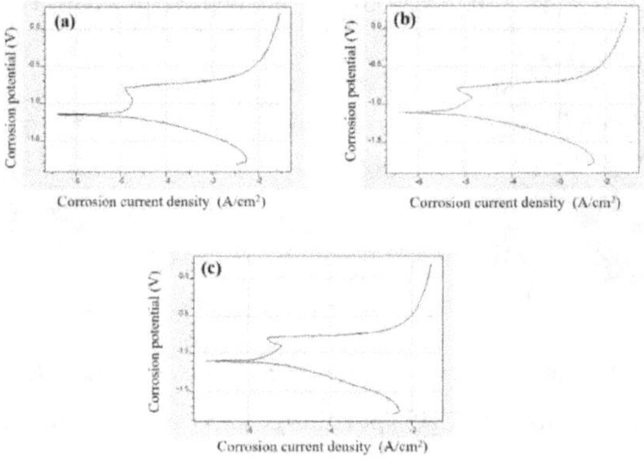

Fig.5.22 Tafel plots of AA7075 surface nanocomposites (a) AA7075+GO (b) AA7075+SiC (c) AA7075+Gr

The behavior of corrosion differs from the microstructural features of materials in which the finer grains produced higher corrosion rates than coarser grains (Feng et al. 2022; Ralston et al. 2010). As observed through OM and SEM images, Gr reinforcement showed UFG structure which enhances grain boundary (GB) density, which exposes more reactive sites to the medium of corrosion, which in turn accelerates localized corrosion. Localized corrosion likely happens when the E_{corr} is minimum (Karmuhilan and Somasundaram 2023). The relationship between corrosion potential (E_{corr}) and corrosion current density (i_{corr}) for the corroded samples is shown in Fig.5.20 (a-c), and their values are extracted based on the Tafel method (Ge et al. 2022). Tafel plots for AA7075 reinforced with GO, SiC and Gr are shown in Fig.5.22 (a), Fig.5.22 (b) and Fig.5.23 (c), respectively. E_{corr} (V) and I_{corr} (A/cm^2) values are tabulated in Table 5.14. As evidenced from Tafel plot, the Gr reinforced surface exhibits more localized corrosion since the E_{corr} value is lowest (-1.224 ± 0.053 V) and I_{corr} is highest (0.000191 ± 0.000024 A/cm^2), yielding with its least corrosion rate (0.966 ± 0.11 mm/year).

Table 5.14 Corrosion measurements of AA7075 surface nanocomposite

Combination	E_{corr} (V)			I_{corr} (A/cm^2)		
	Mean	SD	SE	Mean	SD	SE
AA7075+GO	-1.117	0.086	0.050	0.000044	0.000009	0.0000052
AA7075+SiC	-1.097	0.024	0.014	0.000022	0.000003	0.0000017
AA7075+Gr	-1.224	0.053	0.031	0.000191	0.000024	0.0000140

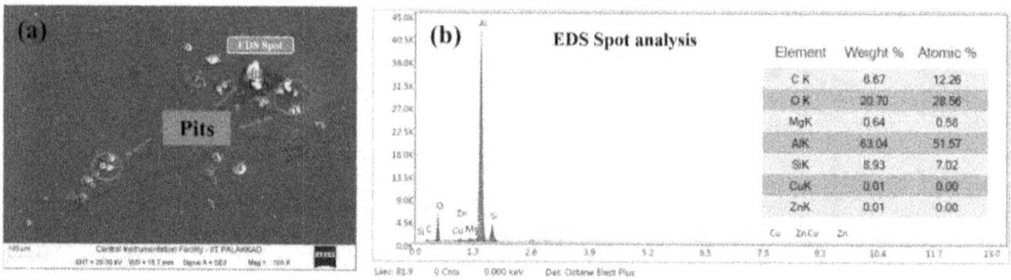

Fig.5.23 FESEM image of corroded AA7075+GO surface (a) EDS spot on corroded product (b) EDS spot analysis on corroded product

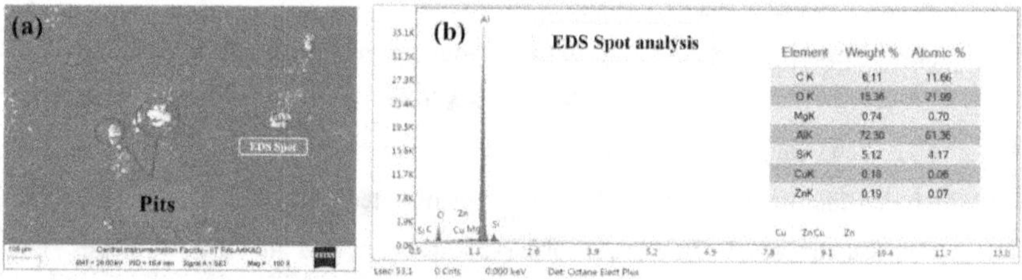

Fig.5.24 FESEM image of corroded AA7075+SiC surface (a) EDS spot on corroded product (b) EDS spot analysis on corroded product

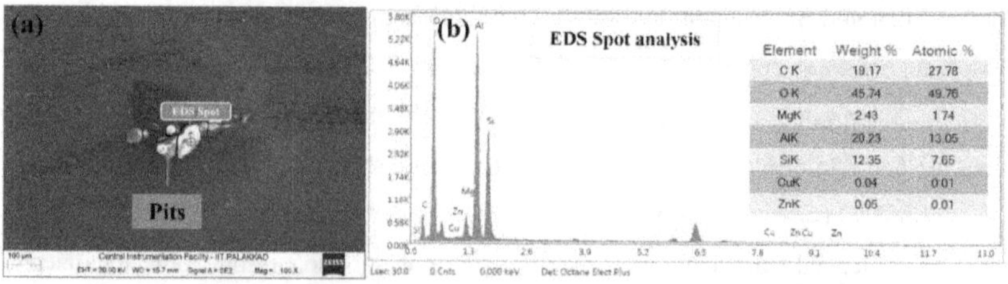

Fig.5.25 FESEM image of corroded AA7075+Gr surface (a) EDS spot on corroded product (b) EDS spot analysis on corroded product

The corroded surface morphology of surface nanocomposites was examined through FESEM-EDS analysis, it as shown in Fig.5.23 to Fig.5.25. All corroded surface exhibited localized pitting corrosion, but with distinct variation in surface morphology and elemental composition of the regions of corrosion.

The AA7075+GO corroded surface revealed (Fig.5.23 a-b) moderate pitting with a significant existence of oxygen (O) and carbon (C) of 20.70 wt. % and 6.67 wt. %, respectively. The more amount of Aluminium (Al) contribution (63.04 wt. %) suggests partial surface degradation. The corrosion performance of GO reinforced surface lies between SiC and Gr reinforced surfaces. The GO reinforced surface exhibited reduced corrosion rate, owing the presence of oxygen-containing functional groups in GO enhances its dispersion and interfacial interaction with the matrix. In contrast, the AA7075+SiC surface showed (Fig.5.24 a-b) relatively deeper pits and higher Al content (72.30 wt. %), with slightly lower O (15.36 wt. %) and C (6.11 wt. %). SiC reinforcement reduced corrosion rate in AA7075 surfaces by enhancing passive film formation, and acting as a barrier to corrosive agents. Uniformly dispersed SiC nanoparticles reduced localized corrosion by stabilizing the microstructure and declining galvanic activity (Aperador et al. 2024; Marode et al. 2023). As shown in SEM image, corroded surface of SiC reinforced AA7075 was comparatively devoid of defects with minimum corroded products owing to the fact that even distribution of particles protecting the surface from corrosion (Bharat and Bose 2023). The AA7075+Gr surface (Fig.5.25 a-b) exhibited the most severe corrosion features, with a thick corrosion layer and intense pit formation.

The EDS analysis confirms very high O and C, and significantly lower Al content. This reveals aggressive localized corrosion, likely accelerated by the cathodic behaviour of Gr, which promotes galvanic coupling with the aluminium matrix. The corroded surface topography of AA7075+GO surface (Fig.5.26 a-c) shows a relatively smoother with less aggressive pit formation. The surface roughness plot of a corresponding pit shows shallow depth and uniform geometry of pit, revealing that the addition of GO particles contributes to reduced corrosion rate (Fig.5.26 a). The surface irregularities of a corroded pit is 2.140 μm. Fig.5.26 (b) shows 3D topographical view of overall corroded surface, with height variations observed between 0 and 301.2 μm. A magnified view of a detailed corroded pit is shown in Fig.5.26 (c), reveals the height difference within the pit is upto 0.20 mm, indicating reduced pit density and smaller pit diameters, reveals a more uniform corrosion mechanism. Fig.5.27 (a-c) shows the surface topography of corroded AA7075+SiC. As shown in Fig.5.27 (a), R_a of a corroded pit is measured at 1.812 μm, which shows lower than that of AA7075+GO surface. Figure 5.27 (b) shows 3D topographical view of the overall corroded surface, revealing height variations observed from 0 to 266.9 μm, indicating less than the height observed in the AA7075+GO.

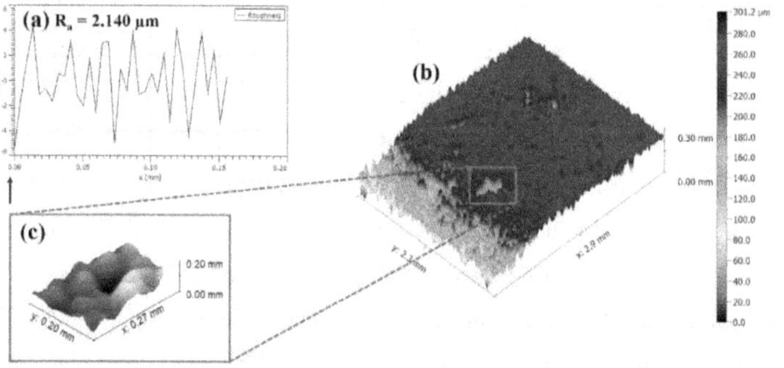

Fig.5.26 Surface topography images of corroded AA7075+GO (a) Surface roughness plot of a pit (b) Overall corroded surface (c) Magnified view of a pit presence in a corroded surface

A detailed view of a corroded pit is shown in Fig.5.27 (c) visualize the height difference of 0.18 mm, showing less degradation than GO based reinforcement. Figure 5.28 (a-c) shows the AA7075+Gr surface topography. As shown in Fig.5.28 (a), R_a of a corroded pit is measured at 2.325 μm, displays the most severe surface damage among other surfaces. It further reveals deep and jagged pits with high aspect ratios. Fig.5.28 (b) shows 3D topographical view of overall corroded surface, with height variations observed between 0 and 348.5 μm. Through galvanic coupling, graphite, which is cathodic to aluminium and electrically conductive, can accelerate corrosion. Extensive pitting, with interconnected pit structures and substantial surface material loss, is visible in the magnified pit image, as shown in Fig.5.28 (c). These characteristics imply that graphite encourages aggressive localized corrosion, which lowers the composite's overall corrosion resistance. Comparison of surface topography among corroded samples are displayed in Table 5.15. AA7075+SiC produced the least R_a values (both maximum and pitting), shallower pits, a smoother corroded surface morphology and the best corrosion resistance. In contrast, the AA7075+Gr exhibited severe corrosion damage with the highest surface roughness and deepest pits. The corroded surface morphology of AA7075+GO lies between others

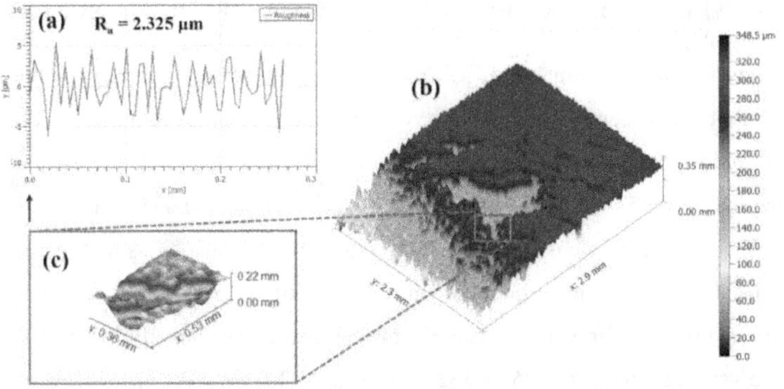

Fig.5.27 Surface topography images of corroded AA7075+SiC (a) Surface roughness plot of a pit (b) Overall corroded surface (c) Magnified view of a pit presence in a corroded surface

Table 5.15 Comparison of surface topography among corroded samples

Combination	Average roughness (R_a)		Pit nature	Surface nature	Corrosion resistance
	Maximum	Pitting			
AA7075+GO	301.2 µm	2.140 µm	Moderate	Rough	Moderate
AA7075+SiC	266.9 µm	1.812 µm	Shallower	Less rough	Best
AA7075+Gr	348.5 µm	2.325 µm	Deep	Very rough	Least

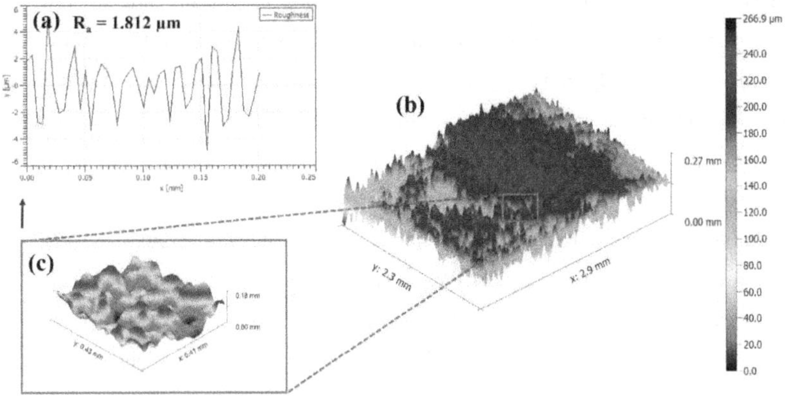

Fig.5.28 Surface topography images of corroded AA7075+Gr (a) Surface roughness plot of a pit (b) Overall corroded surface (c) Magnified view of a pit presence in a corroded surface

5.4 COMPARISON OF CORROSION BEHAVIOR OF AA7075 SURFACE NANOCOMPOSITES

The corrosion performance of AA7075 alloy under various operating conditions show a trade-off between mechanical improvement and corrosion resistance. AA7075 showed a minimum corrosion rate of 0.063 mm/year, revealing better corrosion resistance than FSPed AA7075, which showed a higher rate of 0.095 mm/year at 25 mm/min TTS. This enhance is attributed to precipitate coarsening and solute-depleted zones caused by FSP, which promote corrosion initiation (Kumar et al. 2017b). Similarly, in Friction Stir Welding (FSW), the SZ had a significantly higher CR of 3.128 mm/year than the AA7075 (1.3325 mm/year). The grain refinement owing to Severe Plastic Deformation (SPD) and heat input enhances surface reactivity and corrosion susceptibility (Prabhuraj et al. 2025).

In contrast, the current research exhibited that the CR of fabricated AA7075 surface nanocomposites processed at 30 mm/min of TTS using a plain cylindrical tool pin geometry ranged between 0.0264 and 0.0525 mm/year. This shows an improvement in corrosion performance compared to earlier works on AA7075 and unreinforced FSPed AA7075. Similarly, AA7075+MoS_2/CeO_2 surface composites processing at 30 mm/min of TTS increased corrosion resistance. The presence of CeO_2 contributed to the formation of a stable passive film, leading to an 11% enhance in corrosion potential and a 90% decline in corrosion current density. Although MoS_2 introduced minor micro galvanic effects, the hybrid composite showed better corrosion resistance, highlighting the effectiveness of reinforcement strategies in mitigating corrosion in processed alloys (Maji et al. 2021). The difference between objective 1 and 2 of processed AA7075 surface nanocomposites (Table 5.16) highlights distinct variances in surface morphology, corrosion behavior, and surface topography, all of which are impacted by the TP profile and processing conditions.

In objective one, utilizing a plain cylindrical TP profile with variable TTS, the CS ranged from 47 nm to 66 nm, and the ε values differed from 0.0283 to 0.0391. These microstructural features resulted in relatively lower CR (0.00134 to 0.0264 mm/year), demonstrating better corrosion resistance. Surface morphology observations showed moderate pitting, particularly in the SiC+25 combination, while surface roughness values revealed a smoother finish compared to second objective. The surface topography also demonstrated better overall corrosion resistance, with less severe degradation. In contrast, objective two processed with a threaded cylindrical TP profile, fixed processing parameters and unique reinforcement combinations, resulted in slightly smaller CS (ranging from 45 nm to 56 nm) and lower ε (0.0330 to 0.0404). However, these samples showed significantly higher CR (0.0633 to 0.283 mm/year).

The surface morphology showed deeper pitting and more severe oxidation, especially in the Gr+30 and SiC+30 combinations, which revealed higher oxygen enrichment. Surface roughness values on corroded surfaces were notably larger, indicating that these surface nanocomposites had more substantial surface degradation. Overall, objective one showed better corrosion resistance than objective two. Objective two demonstrated higher CR and greater material degradation owing to larger surface roughness and aggressive oxidation, suggesting that the threaded pin profile, fixed processing variables and unique reinforcements may exacerbate corrosion susceptibility.

Table 5.16 Comparison of the corrosion behavior of AA7075 surface nanoc

Objective	Combination	CS (nm)	ε	CR (mm/year)	Surface morphol observations
One	GNPs+20	56.850 ± 0.900	0.033 ± 0.009	0.0314±0.0325	Pits are less severe SiC+25, but more t GO+30. Corroded su showed moderate pi with oxygen contami of 49.43 wt. %
	SiC+25	47.162 ± 0.054	0.039 ± 0.010	0.0525±0.0237	Pits are severe and la unlike SiC+25 and G indicating severe pi corrosion. Also supp the oxygen (54.66 w dominant at corroded
	GO+30	66.442 ± 0.998	0.028 ± 0.009	0.0264±0.0150	Pits are less aggressiv shallow than SiC+25 GNPs+20. Oxygen (4 wt. %) contaminatio less at corroded surf

Table 5.16 Comparison of the corrosion behavior of AA7075 surface nanocomposite

Objective	Combination	CS (nm)	ε	CR (mm/year)	Surface morphology observations
Two	Gr+30	45.212 ± 0.522	0.040±0.0015	0.9663±0.1098	Severe pitting induced corrosion with oxygen contamination of 45.74 wt. % indicated significa material degradation
	SiC+30	56.235 ± 2.058	0.033±0.0021	0.1077±0.0070	Pits are less aggressive with minimum oxyger (15.36 wt. %) and maximum aluminium (72.30 wt. %) contamination at corrod zone
	GO+30	52.972 ± 3.465	0.036±0.0020	0.2830±0.0151	Moderate pitting with existence of 20.70 wt. % oxygen. More amount aluminum element presents at corroded zo indicated less material degradation

Table 5.17 compares the present findings of the corrosion behaviour of AA7075 surface nanocomposites with other surface nanocomposites processed through FSP. The corrosion behaviour was influenced by workpiece condition (annealed, unreinforced, and heat treated), TRS, TTS, grain size, grain refinement, reinforcement (nature, distribution and amount it contains), tool pin profile and a number of FSP passes. The AA7075+GO sample shows the best corrosion resistance among all the composites under study, with the lowest corrosion current density (7.20×10^{-6} A/cm^2) and corrosion rate (0.0297 mm/year). The addition of GO greatly enhanced the overall electrochemical behaviour, even though its corrosion potential (E_{corr}) (-1.0207 ± 0.0016 V) was more negative than that of the Ti+Al_2O_3 combination. The AA7075+SiC composite in the current research showed a much lower corrosion current density (i_{corr}) (1.053×10^{-7} A/cm^2) and CR (0.0525 mm/year) than the AA7075+SiC corroded surface the same traverse speed of 25 mm/min, as reported elsewhere (Kumar et al. 2017a). This implies that the changes in TRS influenced the CR. The corrosion resistance of the AA6061+ Silicon oxide (SiO_2) surface was reduced due to finer grains as an effect of SPD (Mazaheri et al. 2019). This work aligns with current findings that the Gr+30 combination showed refined microstructure with an average CS of 45.212 nm, leading to an increased CR, unlike 66.442 ± 0.998 nm. The fabricated AA7075+GO nanocomposite demonstrated its potential for high-performance applications in corrosive conditions by outperforming the majority of current FSPed systems overall.

Table 5.17 Comparison of the corrosion behavior of FSPed surface nanocom

Reference	Surface nanocomposite	TRS (rpm)	TTS (mm/min)	E_{corr}	i_{corr}	CR	
(Najar et al. 2025)	AA6082+SiC	900	40	-	-	0.0451 mm/year	
(Kumar et al. 2017a)	AA7075+SiC	1025	25	−950 ± 5mV	2.52 ± 1.2 μA/cm^2	-	i
(Kumar et al. 2023)	AZ31+ Alumina (Al_2O_3)	900, 1200 and 1500	40	−1.3370 V	5.690×10^{-5} A/cm^2	-	re
(Alishavandi et al. 2020)	AA1050+Misch metal oxide (MMO)	1600	100	−775 mV	0.7 μA/cm^2	0.29 m/year	
(Vakili-Azghandi et al. 2024)	Titanium+ Al_2O_3	1000	28	−0.1920 V	0.0081 μA/cm^2	-	a re

Table 5.17 Comparison of the corrosion behavior of FSPed surface nanocom

Reference	Surface nanocomposite	TRS (rpm)	TTS (mm/min)	E_{corr}	i_{corr}	CR	
(Vakili-Azghandi et al. 2024)	Titanium+ Al_2O_3	1000	28	−0.1920 V	0.0081 μA/cm^2	-	In c Al_2C ex
(Mazaheri et al. 2019)	AA6061+Silicon oxide (SiO_2)	500	15	−56.99 V	0.6326 μA/cm^2	-	In co resist owi res
(Vijaya Kumar et al. 2015)	AA7075+B_4C	750	40	−630 mV	-	-	Pitting owing cont Re sh
(Gu et al. 2019)	Ti-35Nb-2Ta-3Zr+ Titanium oxide (TiO_2)	650 and 850	50	−424.7 ± 49.10 mV	22.85 ± 3.27 nA/cm^2	-	Inc
(Zhang et al. 2017)	Ti-6Al-4V+TiO_2	375	60	−0.545 V	0.078 μA	-	Cor
(Liu et al. 2021)	Mg-Zn-Ca+Magnesium oxide (MgO)	800, 1000 and 1200	60 and 80	−1.332 V	47.9 μA/cm^2	-	The s an resp

Table 5.17 Comparison of the corrosion behavior of FSPed surface nanocomposites (

Reference	Surface nanocomposite	TRS (rpm)	TTS (mm/min)	E_{corr}	i_{corr}	CR
Present findings	AA7075+SiC	1500	25	−1.0215±0.017 V	$1.053\times10^{5}\pm$ 0.02×10^{-6} A/cm^2	0.0525±0.0237 mm/year
	AA7075+GO	1500	30	−1.0020±0.047 V	$1.030\times10^{5}\pm$ 7.20×10^{-6} A/cm^2	0.0207±0.0150 mm/year
	AA7075+GNPs	1500	20	−1.1675±0.082 V	$1.480\times10^{-5}\pm0.06\times10^{-6}$ A/cm^2	0.0314±0.0325 mm/year
	AA7075+GO	1480	30	−1.117±0.086	$4.4\times10^{-5}\pm$ 9×10^{-6} A/cm^2	0.2830±0.0151 mm/year
	AA7075+SiC	1480	30	−1.097±0.024	$2.2\times10^{-5}\pm$ 3×10^{-6} A/cm^2	0.1077±0.0070 mm/year
	AA7075+Gr	1480	30	−1.224±0.053	$1.91\times10^{-4}\pm$ 2.4×10^{-5} A/cm^2	0.9663±0.1098 mm/year

5.5 PRACTICAL APPLICATIONS OF AA7075 SURFACE NANOCOMPOSITES IN CORROSIVE ENVIRONMENTS

The current findings highlight that the GO+30 employed using plain cylindrical TP profile are exhibited significant corrosion resistance, making the GO reinforced AA7075 surface an ideal candidate material for marine environments, including off-shore structures and shipbuilding. Higher TTS improved reduced corrosion rates, highlighting the significance of process parameter optimization setting marine applications. In contrast, the Gr+30 combination (objective two) showed an enhanced CR owing to refined CS produced by threaded cylindrical TP profile, further resulting in severe pitting corrosion, which makes it less preferable for aquatic environments. GO+30 combination yields finer surface topography and functional groups, increase their resistance to corrosive environments and guarantees their efficacy when exposed to seawater and other comparable corrosive agents over an extended period.

Comparative analysis in Table 5.17 and Table 5.18 demonstartes GO reinforced aluminium alloys undergo FSP, significantly improving corrosion resistance. Although corrosion benchmarking is usually conducted in maritime environments, GO+30 superior surface behavior indicates broader applicability in various situations, including the chemical, automotive, and aerospace sectors. These surface nanocomposites are appropriate for applications needing strength and durability in corrosive or high-stress environments because of their twofold improvement, which is accomplished by grain refinement and nano-reinforcement. GO provides a scalable and economical substitute for conventional reinforcements, allowing for the mass market adoption of corrosion-resistant materials at lower costs. Regarding sustainability, the enhanced passivation behavior lessens the need for ecologically damaging coatings, which supports more environmentally friendly engineering solutions in building and infrastructure. Finally, these surface nanocomposites' qualities make them excellent options for next-generation materials in harsh settings where oxidation, radiation, and temperature cycling are significant issues, such as deep-sea and aerospace missions.

5.6 SUMMARY

The chapter investigated the corrosion behaviour of AA7075 surface nanocomposites through potentio dynamic polarization test. Statistical analysis including descriptive statistics, normality, non-parametric tests, and ANOVA were applied to interpret the corrosion rate data. The corroded samples were further examined through surface morphology and topography analyses to understand corrosion behaviour. The summary points are as follows.

- ANOVA results confirmed that reinforcement nature significantly influence the corrosion rate while employed using threaded tool profile, unlike cylindrical tool profile.
- Grain refinement increased grain boundary density, which enhanced electrochemical reactivity and reduced corrosion resistance, particularly in highly refined structures. Under threaded tool pin profile, reinforcement material became the dominant corrosion controlling factor.
- GO+30 exhibited lower corrosion current density, indicating outstanding corrosion resistance. GO reinforcement effectively reduced pit intensity and depth, as its functional groups helped stabilize the matrix and suppress electrochemical attack.
- Surface morphology showed carbon-reinforced AA7075 surface nanocomposites developed less aggressive and shallower pits compared to ceramic reinforcement systems.
- Graphene reinforcement triggered cathodic behavior and galvanic effects, leading to severe localized corrosion and oxygen-rich degradation regions.
- SiC reinforcement produced lower surface roughness and shallower pits, confirming its protective role against localized corrosion.

CHAPTER 6

TRIBOLOGY BEHAVIOUR OF AA7075 SURFACE NANOCOMPOSITES

6.1 INTRODUCTION

This chapter investigates the wear behavior of AA7075 surface nanocomposites under room and elevated temperature dry sliding wear conditions. A pin-on-disc tribometer was employed to evaluate the tribological behaviour of AA7075 surface nanocomposites. The tribology behaviour of AA7075 surface nanocomposites was structured around two main objectives: first, to study the influence of Tool Traverse Speed (TTS) on Specific Wear Rate (SWR), and second, to study the impact of specific nano reinforcement combinations (AA7075+GO, AA7075+SiC, and AA7075+GNPs) on SWR. The SWR was examined statistically using descriptive statistics (mean, standard deviation and variance), normality tests, non-parametric tests, and Analysis of Variance (ANOVA). Based on statistical data analysis, the worn out samples were interpreted using microhardness, surface morphology and topography techniques.

6.2 DRY SLIDING TRIBOLOGY BEHAVIOR UNDER ROOM TEMPERATURE CONDITION

AA7075 surface nanocomposites were friction stir processed (FSPed) at varying TTS, resulting in nine combination of experiements. To find the influence of TTS on SWR, a dry sliding tribology behaviour were studied under room temperature testing condition. The applied load, sliding velocity, and sliding speed of 20 N, 2 m/s, and 383 rpm, respectively, were used as wear test parameters. The SWR was calculated on each AA7075 surface nanocomposites and average of three measurements were recorded to ensure accuracy. The obtained SWR data were analysed statistically including descriptive statistics, normality study, homogeneity of variance and ANOVA. Based on the statistical outcomes, selected worn samples were further analyzed through microhardness, surface morphology and topographical characterization techniques to reveal the underlying wear mechanisms.

6.2.1 Statistical analysis of the SWR

Table 6.1 displays the descriptive statistical measures including mean, median, standard deviation (SD) and variance (Var) of the SWR of AA7075 surface nanocomposites under room temperature dry sliding wear behavior condition. Table 6.1 show that the best combination is GO at 25 mm/min, which produces the lowest mean SWR (0.000127 mm^3/Nm) with the minimum amount of variability (standard deviation = 0.0000316). This implies that GO reinforcement at this TTS offers consistent performance and better wear resistance. In contrast, GNPs+20 combination was observed as the least wear resistance, exhibiting the highest average SWR of 0.000255 mm^3/Nm and significant variability with standard deviation of 0.0001136, indicating non-consisting tribological behavior. The mean SWR for GO combinations increased from 0.000194 mm^3/Nm at 20 mm/min to 0.000254 mm^3/Nm at 30 mm/min, indicating a decline in wear performance as TTS hiked. With standard deviations ranging from 0.000316 to 0.000698, the variability in the GO combinations remained small across TTS levels, indicating that GO offers wear performance that is balanced but not outstanding. In comparison to SiC and GNPs, this behavior demonstrates the intermediate potential of GO as a reinforcement, providing moderate wear resistance and consistency.

Table 6.1 Descriptive statistical analysis of the SWR of AA7075 surface nanocomposite

Combination	SWR (mm^3/Nm) ($\times10^{-4}$)			Mean ($\times10^{-4}$)	Median ($\times10^{-4}$)	SD ($\times10^{-4}$)	Var ($\times10^{-9}$)
	R_1	R_2	R_3				
GO+20	2.72	2.14	1.33	2.06	2.14	0.698	4.9
GO+25	1.49	1.41	0.91	1.27	1.41	0.316	1.0
GO+30	2.62	2.38	1.58	2.19	2.38	0.540	2.9
SiC+20	2.45	1.45	1.92	1.94	1.92	0.502	2.5
SiC+25	1.66	1.12	1.59	1.46	1.59	0.294	0.8
SiC+30	3.76	1.49	2.38	2.54	2.38	1.139	12.8
GNPs+20	3.72	1.45	2.47	2.55	2.47	1.136	12.9
GNPs+25	2.80	1.04	0.93	1.59	1.04	1.053	11.0
GNPs+30	2.78	3.19	0.98	2.32	2.78	1.173	13.7

Variability analysis showed that SiC generally exhibited the least fluctuation across low and moderate TTS levels. GNPs showed a substantial degree of variability, especially at lower TTS, exhibited inconsistent wear performance. These results highlight how important TTS and reinforcement type are in influencing wear behavior. The most efficient combination was GO at 25 mm/min; however, SiC showed consistent improvement as TTS increased, and GNPs showed notable fluctuation, which limited its efficacy. Table 6.2 displays the mean SWR across RT and TTS. Among the RT, GO exhibited lower mean SWR of 0.0001842 mm^3/Nm with least SD of 0.00006387, unlike GNPs recorded a mean SWR of 0.0002151 mm^3/Nm with highest SD of 0.000106346. SiC with a mean SWR 0.0001980 mm^3/Nm with SD of

0.00007959, falls between GO and GNPs. Regarding TTS, 25 mm/min resulted in the lowest mean SWR of 0.0001439 mm^3/Nm with least SD 0.0000586, while 30 mm/min produced the highest mean wear rate of 0.0002351 mm^3/Nm with highest SD of 0.0084. The wear performance at 20 mm/min, with an average SWR of 0.0000765 mm^3/Nm with SD of 0.0000765, was intermediate. Among the group combinations (RT and TTS), GO+25 demonstrated the lowest mean SWR, unlike GNPs+30. The mean SWR of SiC+20 lied between GO+25 and GNPs+30.

Table 6.2 Mean SWR ($\times10^{-4}$) in mm^3/Nm of AA7075 surface nanocomposites across RT and TTS

RT	20 mm/min			25 mm/min			30 mm/min			Mean	SD
GO	2.72	2.14	1.33	1.49	1.41	0.91	2.62	2.38	1.58	1.842	0.6387
SiC	2.45	1.45	1.92	1.66	1.12	1.59	3.76	1.49	2.38	1.980	0.7959
GNPs	3.72	1.45	2.47	2.80	1.04	0.93	2.78	3.19	0.98	2.151	0.1063
Mean	2.183			1.439			2.351			-	-
SD	0.765			0.586			0.875			-	-

Table 6.3 Shapiro-Wilk test of the SWR of AA7075 surface nanocomposite

Combination	W	Critical observations from statistical table		
		Critical values	Confidence levels	Normality status
GO+20	0.991	0.959 to 0.998	$P_{0.5}$ to $P_{0.9}$	Data follows normal distribution
GO+25	0.852	0.789 to 0.959	$P_{0.1}$ to $P_{0.5}$	
GO+30	0.912	0.789 to 0.959	$P_{0.1}$ to $P_{0.5}$	
SiC+20	0.999	0.999	$P_{0.95}$	
SiC+25	0.840	0.789 to 0.959	$P_{0.1}$ to $P_{0.5}$	
SiC+30	0.984	0.959 to 0.998	$P_{0.5}$ to $P_{0.9}$	
GNPs+20	0.996	0.959 to 0.998	$P_{0.5}$ to $P_{0.9}$	
GNPs+25	0.795	0.789 to 0.959	$P_{0.1}$ to $P_{0.5}$	
GNPs+30	0.885	0.789 to 0.959	$P_{0.1}$ to $P_{0.5}$	

The Quantitle Quantile (Q-Q) plots for different combinations are shown in Fig.6.1 (a-i). It shows the alignment of ordered values with theoretical quantiles, indicating normal data distribution for all combinations. To validate Q-Q plots, the Shapiro-Wilk test was conducted. All experimental combinations, were found to be normally distributed by the Shapiro-Wilk test, agrees with Q-Q plots. Table 6.3 displays the Shapiro-Wilk test results on SWR of AA7075 surface nanocomposites.

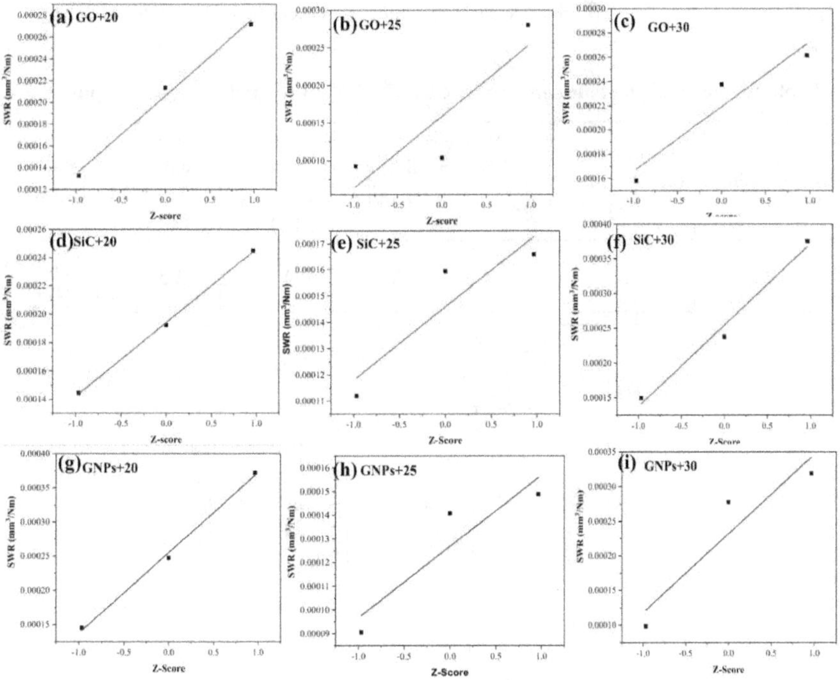

Fig.6.1 (a-i) Q-Q plots of the SWR of AA7075 surface nanoocmposites

There was no discernible variation in variances between the combinations, according to Levene's test for homogeneity of variances ($L = 0.453$, $p > 0.05$). As a result, the ANOVA's assumption of equal variances was maintained. Table 6.4 displays the Levene's test results on SWR. Box plots of the SWR of AA7075 surface nanocomposites are shown in Fig.6.2. At lower TTS, all surface nanocomposites exhibited moderate wear resistance, but the Interquartile range (IQR) was particularly wider for GNPs-based combinations, revealing more variability in their wear performance. In contrast, medium TTS resulted in the lowest and most consistent SWR values, notably for GO+25 and SiC+25, suggesting superior wear resistance. This optimal TTS likely facilitates better dispersion of the reinforcement particles, leading to enhanced wear performance. At higher TTS, the SWR enhanced significantly for SiC+30 and GNPs+30, highlighting that higher TTS cause poor particle distribution and agglomeration, which degrade wear resistance due to weak bonding at the composite-matrix interface. SiC based combinations exhibited more stable performance across TTS, their wear resistance was declined compared to GO at medium TTS.

Table 6.4 Levene's test outcomes of the SWR of AA7075 surface nanocomposite

No. of combinations (k)	Total no. of observations (N)	Degrees of freedom (df)		Calculated test statistics, (L)	Critical value from F-table	Test status
		$df_1=N-k$	$df_2=k-1$			
9	27	18	8	0.453	2.51	Fail to reject H_0

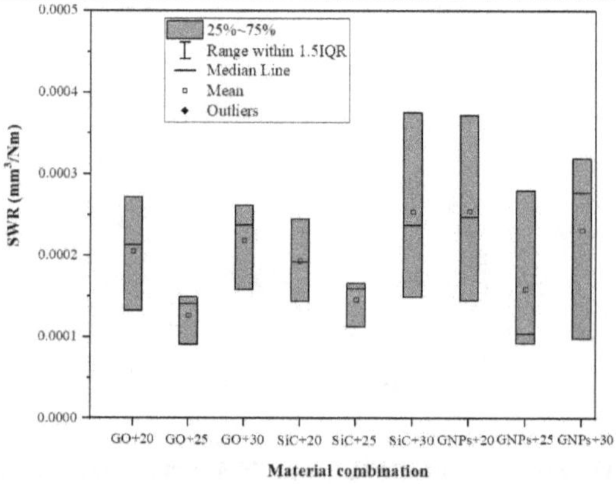

Fig.6.2 Box plot visualization of the SWR of AA7075 surface nanocomposite

The narrowest IQRs observed at medium TTS, notably for SiC and GO reinforcements, exhibit that this TTS not only provides the best wear resistance but also ensures more consistent and reliable tribological performance. This finding reveals the importance of optimizing processing parameters, notably TTS, in increasing the wear resistance of FSPed surface nanocomposites. After validating the results obtained from normality and homogeneity of variance, ANOVA is being used to compare the combination of means. According to ANOVA reports ($F < F_{critcal}$), none of those factors significantly affect SWR. The ANOVA for SWR is displayed in Table 6.5. The observed differences among combinations are likely due to statistical noise (random fluctuations) rather than true variations in material characteristics. In addition, it may indicate that other factors, such as environment conditions or microstructural characteristics could be more dominant in determining the SWR.

Table 6.5 ANOVA outcomes of the SWR of AA7075 surface nanocomposite

Source of variation	Sum square (SS) ($\times 10^{-8}$)	df	Mean square (MS) ($\times 10^{-8}$)	Calculated, F	Critical value from F-Table	Test status
TTS	4.24	2	2.120	2.825	3.55	No significant effect
RT	0.04	2	0.020	0.030		
TTS*RT	0.93	4	0.234	0.309	2.93	
Error	13.51	18	0.750	-		
Total	18.72	26	-			

6.2.2 Worn surface analysis

The effect of micro hardness on SWR and Coefficient of Friction (COF) of AA7075 surface nanocomposites is shown in Fig.6.3. It indicates that the both reinforcement and TTS are significantly influencing tribological characteristics. GO+25 performed well in terms of SWR, unlike SiC+25 and GNPs+25. GO+25 demonstrated improved resistance to deformation by achieving the hardness of 130 HV due to solid lubricant and easy shear between layers, reducing friction. Grain refinement through plastic deformation and Dyanmic Recrystallization (DRX) enhanced the hardness of the FSPed AA7075 surface, consistent with the Hall–Petch relation, where finer grains yield higher hardness. Incorporating GO particles further enhanced hardness owing to thermal mismatch between reinforcements and AA7075-T6 alloy, pinning of Grain Boundaries (GBs), dispersion effects, dislocation blocking through Orowan strengthening and strain field mismatch between GO and the matrix (Sudhakar et al. 2016). The efficacy of GO was also validated by the wear behaviour as determined by SWR. This sample showed the lowest SWR of 0.000127 mm^3/Nm, indicating that GO reinforcement offers better protection against material loss under sliding conditions. With a hardness of 127 HV and a SWR of 0.000194 mm^3/Nm, the SiC+20 demonstrated substantial improvements, while the GNPs+30 showed the least outcomes, with a hardness of 117 HV and a SWR of 0.000232 mm^3/Nm. This could limit the efficiency of the reinforcement owing to particle aggregation and poor bonding at higher TTS.

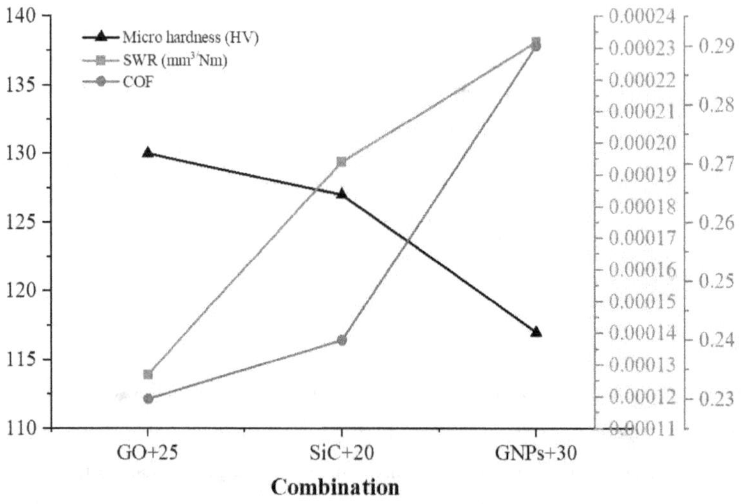

Fig.6.3 Micro hardness versus SWR and COF of AA7075 surface nanocomposites

A similar pattern was observed in frictional behaviour. With the lowest coefficient of friction (0.23), the GO sample demonstrated less energy loss and smoother sliding interactions. The GNPs sample enhanced the COF of 0.29, presumably as a result of surface imperfections and unequal distribution, whereas the SiC sample displayed a slightly higher COF of 0.24. Based on outcomes, GO reinforcement greatly improves the hardness, wear resistance, and COF of AA7075 when processed at an ideal TTS of 25 mm/min, a viable choice for parts used in automotive and aerospace industries where high strength and low wear are essential. SiC provides moderate advantages in the meantime, whereas GNPs at higher TTS exhibit drawbacks that needs more research to enhance bonding and dispersion. The surface morphologies of worn out surfaces are shown in Fig.6.4 (a-c). It shows that the both reinforcement and TTS are significantly influencing surface morphologies. GO+25 surface (Fig.6.4 a) shows excellent resistance to wear with consistent grooves and a minimum debris, indicating abrasive type wear mechanism (Michael Rajan et al. 2014). While the moderate TTS ensured optimal heat input and dispersion of the reinforcement, leading to a refined microstructure with no flaws, the GO reinforcement probably helped to improve load-bearing capacity and decrease frictional interactions.

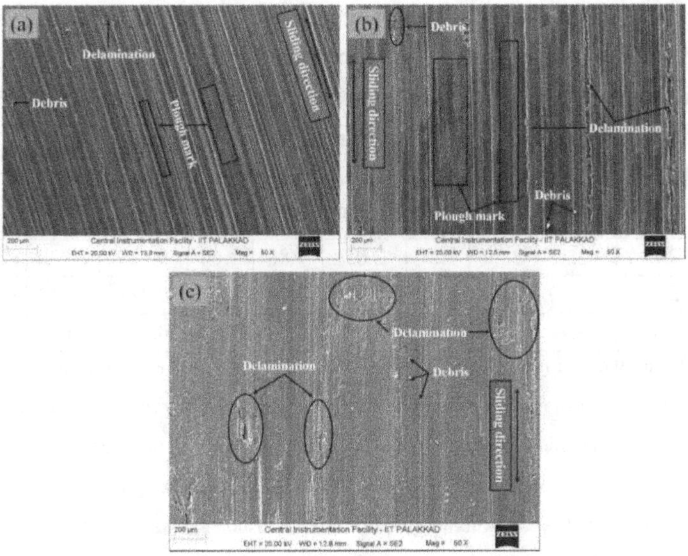

Fig.6.4 SEM morphology of worn surfaces under room temperature dry sliding wear condition (a) GO+25 (b) SiC+20 (c) GNPs+30

In contrast to GO+25, the worn grooves are deeper and wider for SiC+20, and there is a discernible build-up of debris and it is shown in Fig.6.4 (b). The TTS of 20 mm/min may have caused localized warming and particle clustering, which could have contributed to stress concentrations and nature of wear mechanism is adhesive (Patil et al. 2022). As shown in Fig.6.4 (c), widespread cracks, delamination, and grooves show severe wear and the concerned wear mechanism is the combined abrasive and adhesive (Gajakosh et al. 2023). The matrix-reinforcement interface may have been weakened by inadequate heat generation and poor particle distribution brought on by TTS of 30 mm/min. Inadequate dispersion and bonding may have increased adhesive and fatigue wear mechanisms under sliding, even though GNPs have great lubricating qualities.The wear tracks details of worn AA7075 surface nanocomposites are displayed in Table 6.6 and it shows GO+25 combination exhibited lowest wear depth, wear width and wear volume, among other combinations. Figure 6.5 (a-b) illustrates the GO+25 worn details . The R_a profile in Fig.6.5 (a) shows minimal wear, with only slight topographical variation along the wear track. The surface indicates a relatively uniform and smooth texture, revealing that GO nanoparticles were effectively enhanced the wear resistance of the AA7075 matrix. The 3D profile shows that the depth and width wear were notably smaller, unlike other worn out surfaces, with the wear depth and width measuring only −0.0013 mm and 0.0974, respectively.

Table 6.6 Details of the wear track of worn AA7075 surface nanocomposites

Combination	Wear depth (mm) ($\times10^{-3}$)			Wear width (mm) ($\times10^{-1}$)			Wear volume (mm^3)		
	Mean	**SD**	**SE**	**Mean**	**SD**	**SE**	**Mean**	**SD**	**SE**
SiC+20	−2.37	2.20	1.27	2.05	1.80	1.04	0.329	0.030	0.0172
GO+25	−1.30	2.22	1.30	0.97	1.69	0.97	0.221	0.013	0.0075
GNPs+30	−3.74	3.24	1.87	5.27	4.57	2.64	0.371	0.003	0.0016

As shown in Fig.6.5 (b), the wear track is narrow and localized, provides a protective layer against material removal. The reduced wear observed can be attributed to the superior mechanical properties of GO, which improve the hardness, lubrication properties and ability to form a protective oxide film, thus minimizing the wear volume. The lowest observed wear volume (0.221 mm^3), further shows the superior wear resistance. Figure 6.6 (a-b) illustrates the SiC+20 worn details, shows more significant wear compared to GO+25, with greater R_a and more pronounced grooves along the wear track. The wear depth and width of −0.0024 mm and 0.2055 mm, respectively, shows larger than those observed in GO+25, indicating that the ceramic particles did not provide as much protection. The enhanced wear observed in Fig.6.6 (a-b) can be attributed by the brittle nature of SiC particles and lacks the inherent toughness of GO. This brittleness makes it more susceptible to fracture under high stress during the dry sliding wear test, leading to enhance material loss. The wear volume (0.329 mm^3) was recorded higher than GO+25, but lesser than GNPs+30.

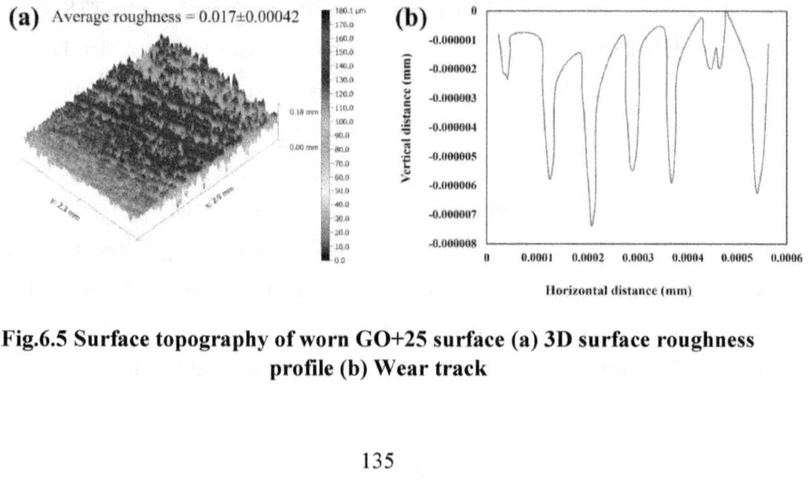

Fig.6.5 Surface topography of worn GO+25 surface (a) 3D surface roughness profile (b) Wear track

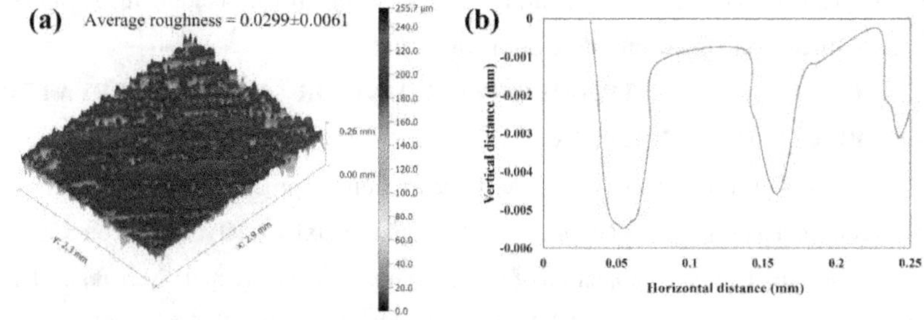

Fig.6.6 Surface topography of worn SiC+20 surface (a) 3D surface roughness profile (b) Wear track

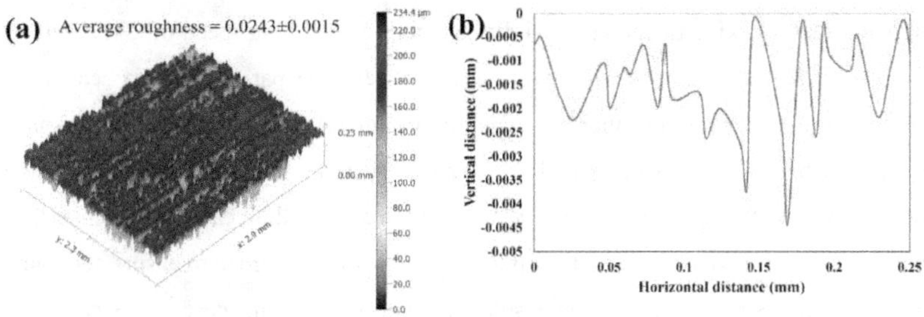

Fig.6.7 Surface topography of worn GNPs+30 surface (a) 3D surface roughness profile (b) Wear track

The moderate wear track and volume can be linked to the relatively high hardness of SiC, which contributes to surface degradation through microcracking and particle detachment. The worn surface of GNPs+30 are illustrated in Fig.6.7 (a-b). It shows a rougher surface profile unlike GO+25 and SiC+20 surfaces, with significant wear marks and deeper grooves. The wear depth of −0.0037 mm and width of 0.5276 mm were the highest among all combinations, and the volume of wear was 0.329 mm^3, matching SiC+20 but still significantly larger than GO+25. The enhanced wear in GNPs+30 can be attributed to two main factors: the relatively high TTS and the agglomeration of GNPs during processing. At higher TTS, the temperature at the tool composite interface surges, potentially leading to the softening of the AA7075 matrix and poor adhesion between the GNPs and the matrix material. The agglomeration of GNPs further decreases the effectiveness of the reinforcement, as large clusters of GNPs are less effective at resisting wear and distributing the load across the surface. As shown in Fig.6.7 (b), the wider and deeper wear track is indicative of these tribological challenges, where the surface is less capable of withstanding the shear and normal forces from the counterface during the wear test.

6.3 DRY SLIDING TRIBOLOGY BEHAVIOR UNDER ELEVATED TEMPERATURE CONDITION

AA7075 surface nanocomposites were friction stir processed (FSPed) at unique nano reinforcements such as Graphene (Gr), Graphene Oxide (GO) and Silicon carbide (SiC) resulting in three combination of experiements. The experiemnts were done at 30 mm/min of TTS and 1480 rpm of TRS, employed by threaded cylindrical tool pin. The aim of this case two is to find the exclusive effect of reinforcement type on elevated temperature dry sliding tribology behaviour of AA7075 surface nanocomposites. The applied load, working temperature, sliding velocity and sliding speed of 20 N, 150^0C, 2 m/s and 383 rpm, respectively, were used as wear test parameters. To accurately represent the operating thermal conditions of AA7075-T6-based aerospace components, including wing panels, fuselage structures, and fastening systems, a temperature of 150°C was chosen. The SWR was calculated on each AA7075 surface nanocomposites and average of three measurements were recorded to ensure accuracy. The obtained SWR data were analysed statistically including descriptive statistics, normality study, homogeneity of variance and ANOVA. Based on the statistical outcomes, selected worn samples were further analyed through microhardness, surface morphology and topographical characterization techniques to reveal the underlying wear mechanisms.

6.3.1 Statistical analysis of the SWR

Table 6.7 displays the descriptive statistical data of the SWR (mm^3/Nm) at 150^0C across different material combinations. For AA7075+GO, the data shows a slight positive skew, as the mean is slightly lesser than the median, indicating that a few tests may have resulted in higher SWR. The moderate SD and variance indicate a moderate spread in the SWR. AA7075+SiC recorded highest mean SWR, SD and variance compared to the other two combinations, suggesting considerable variability in the SWR. This may be uneven distribution or interaction of the SiC particles within the composite material. AA7075+Gr combination showed the lowest mean SWR, SD and variance, indicating more consistent wear behavior unlike other combinations. Q-Q plots of the SWR at 150^0C of AA7075 surface nanocomposites are shown in Fig.6.8 (a-c). It shows that all combinations exhibited closely to the reference line, supporting that the SWR approximately follows normal distribution. To further confirm the normality, Shapiro-Wilk test was conducted.

Table 6.7 Desriptive statistical analysis of the SWR at 150^{0}C across different material combinations

Combination	SWR (mm^3/Nm) ×10^{-4}			Mean	Median	SD	Var
AA7075+GO	2.06	2.59	2.67	2.4×10^{-4}	2.59×10^{-4}	3.3×10^{-5}	1.01×10^{-9}
AA7075+SiC	1.96	2.96	2.76	2.6×10^{-4}	2.76×10^{-4}	5.3×10^{-5}	2.80×10^{-9}
AA7075+Gr	2.21	1.79	1.88	1.9×10^{-4}	1.88×10^{-4}	2.2×10^{-5}	0.49×10^{-9}

Table 6.8 Shapiro-wilk test results of the SWR at 150^{0}C of AA7075 surface nanocomposites

Combination	W	Critical observations from statistical table		
		Critical values	Confidence levels	Normality status
AA7075+GO	0.846	0.789 to 0.959	$P_{0.1}$ to $P_{0.5}$	Data follows normal distribution
AA7075+SiC	0.893	0.789 to 0.959	$P_{0.1}$ to $P_{0.5}$	
AA7075+Gr	0.902	0.789 to 0.959	$P_{0.1}$ to $P_{0.5}$	

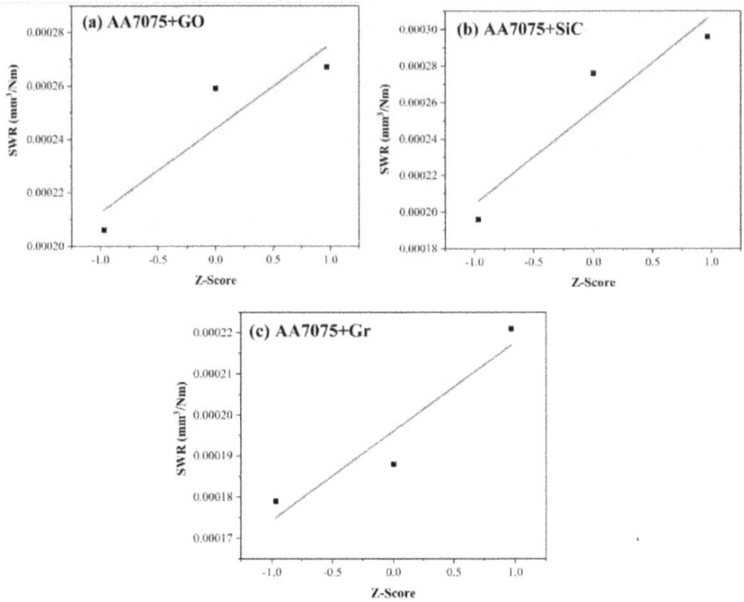

Fig.6.8 (a-c) Q-Q plots of the SWR at 150^{0}C of AA7075 surface nanocomposites

All material combinations, were found to be normally distributed by the Shapiro-Wilk test, indicating that the SWR at 150^0C of AA7075 surface nanocomposites is normally distributed and it is displayed in Table 6.8. This result agrees with Q-Q plots. There was no discernible variation in variances between the combinations, according to Levene's test for homogeneity of variances ($L = 1.382$, $p > 0.05$). As a result, the ANOVA's assumption of equal variances was maintained. Table 6.9 displays the Levene's test results on SWR at 150^0C of AA7075 surface nanocomposites.

Table 6.9 Levene's test outcomes of the SWR of AA7075 surface nanocomposites

No. of combinations (k)	Total no. of observations (N)	Degrees of freedom (df)		Calculated test statistics, (L)	Critical value from F-table	Test status
		df_1=N-k	df_2=k-1			
3	9	6	2	1.382	8.81	Fail to reject H_0

The AA7075+GO combination shows moderate wear resistance with a median SWR of 0.00026 mm^3/Nm and a narrow IQR, suggesting consistent performance, as shown in Fig.6.9. The mean SWR is slightly below the median, indicating a mild left-skewed distribution. The AA7075+SiC combination shows the highest SWR with a larger IQR, revealing significant wear variability. Despite its hardness, SiC's abrasive nature promotes to increased wear. The AA7075+Gr combination exhibits the lowest SWR 0.00018 mm^3/Nm with stable performance. As shown in Table 6.10, the calculated F value is lower than that of critical value, revealing that there is no statistically significant difference among reinforcements in terms of their influence on the SWR. The results indicate that the variation in SWR observed across the various reinforcement combinations is likely due to random factors rather than the particular choice of reinforcement. So, based on this analysis, the type of reinforcement does not have a statistically significant impact on the wear resistance of AA7075 surface nanocomposites under the given experimental conditions.

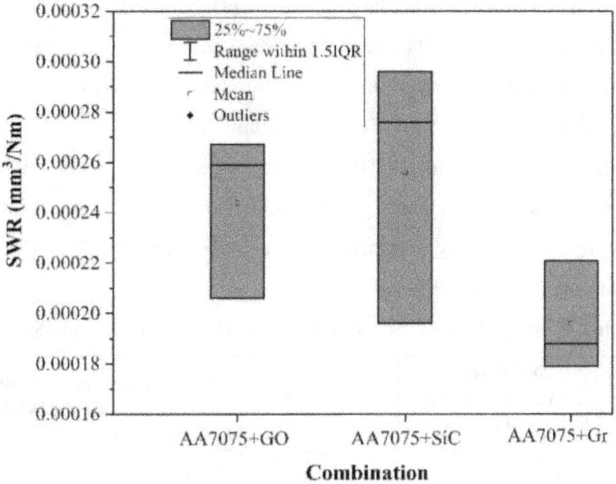

Fig.6.9 Box plot visualization of the SWR at 150^{0}C of AA7075 surface nanocomposites

Table 6.10 ANOVA outcomes of the SWR at 150^{0}C of AA7075 surface nanocomposites

Source of variation	SS	df	MS	Calculated, F	F-Table	Test status
Between SWR	6.0×10^{-9}	2	3.00×10^{-9}	2.412	4.74	Not significant effect on SWR
Within SWR	8.8×10^{-9}	7	1.25×10^{-9}	-		
Total	1.48×10^{-9}	9	-	-	-	-

6.3.2 Worn surface analysis

Table 6.11 displays the maximum processing temperature (PT) during FSP, crystallite size (CS), and compressive residual stress (CRS) of the surface nanocomposites. As shown in Fig.6.10, the highest processing temperature was recorded for AA7075+Gr, followed by AA7075+GO and AA7075+SiC. The temperature variation is associated with the changes in heat generation and thermal conductivity characteristics of the nano reinforcements during processing. The PT is in direct relation with CRS of the surface. AA7075+Gr produced maximum PT, which results enhanced CRS of -128.33 ± 9.07 MPa. The existence of CRS is advantageous for increasing fatigue resistance, and the pronounced effect observed in Gr reinforced surface nanocomposites due to its strong interfacial interaction with the AA7075 matrix, which contributes to greater strain accommodation and stress retention. CS shows inverse relationship with CRS. For a minimum CRS, larger CS was observed. For AA7075+SiC, it produced lower CRS, results largest CS, whereas AA7075+Gr showed the smallest size, revealing a greater refinement effect by Gr owing to its lubricating nature and its ability to restrict grain growth during DRX. AA7075+GO exhibited an intermediate crystallite size.

Table 6.11 PT, CS and CRS of AA7075 surface nanocomposite

Surface nanocomposite	Maximum Temp during FSP	CS (nm)	Compressive residual stress (MPa)
AA7075+GO	336.7 ^{0}C	52.972 ± 3.465	− 96.00 ± 6.24
AA7075+SiC	324.9 ^{0}C	56.235 ± 2.058	−75.33 ± 5.69
AA7075+Gr	347.0 ^{0}C	45.212 ± 0.522	−128.33 ± 9.07

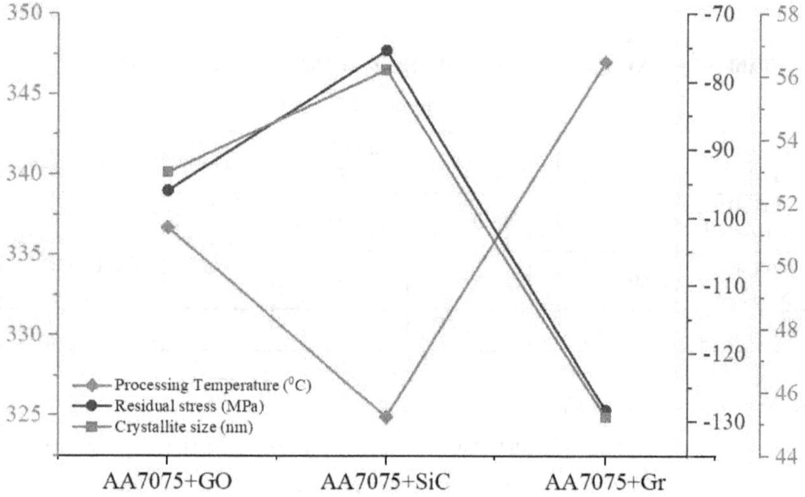

Fig.6.10 PT versus CRS and CS of AA7075 surface nanocomposites

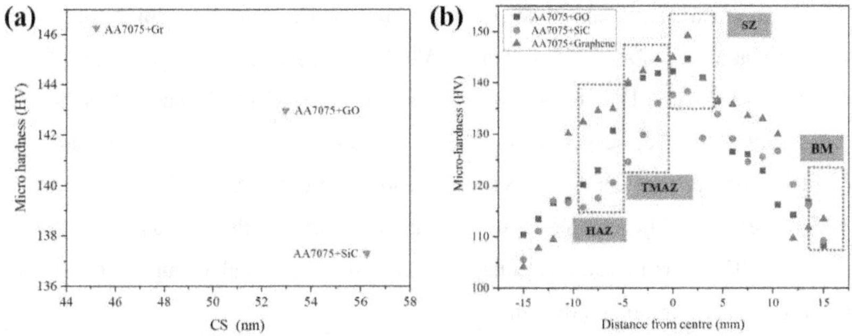

Fig.6.11 CS versus Micro hardness of AA7075 surface nanocomposites (a) Scatter plot shows the influence of CS on HV(b) Variation of micro hardness across zones

Among the fabricated surface nanocomposites, Gr reinforcement exhibited the most significant effect on microstructural refinement and CRS, revealing its potential for enhancing tribological performance of AA7075 surface nanocomposites. Scatter plot shown in Fig.6.11 (a) illustrates the variation in micro hardness with respect to CS. CS shows indirect relation with micro hardness. The maximum micro hardness is achieved, when the CS is minimum. The micro hardness profile of friction stir processed surfaces across different zones are shown in Fig.6.11 (b). The micro hardness is inversely proportional to CS, as shown. The refined grain structure experiences the highest hardness compared to the unaffected zone. The coarse-grained structure of AA7075-T6 alloy exhibited reduced micro hardness, unlike FSPed surface. The average micro hardness of the unaffected zone (BM), HAZ, TMAZ and SZ are 112 HV, 125 HV, 134 HV and 142 HV, respectively, for all surface nanocomposites. As compared with BM, the SZ yields 26.78 % more micro hardness.

The enhancement in micro hardness at SZ observed due to ultra-fine grain refinement on reinforcement addition, homogeneous distribution of reinforcements and strengthening mechanism (Orowan and grain) (Aydin 2021; Jain et al. 2019). The micro hardness starts degrading away from SZ due to precipitate coarsening (Kumar et al. 2024). The mean micro hardness of the AA7075+Gr surface nanocomposite is 146.27 ± 2.55 HV, which results in a 30.59 % improvement with respect to the AA7075 surface. However, the SD is marginally higher than others, the ultrathin, high aspect ratio of Gr may be the cause of this, since it tends to aggregate under mechanical and thermal stresses during FSP. This, in turn, creates localised reinforcement variability, resulting in microstructural non-uniformity and increased hardness variability. The average micro hardness of AA7075+GO and AA7075+SiC surface nanocomposites are 142.97 HV and 137.30 HV, respectively. This improves 27.65 % and 22.58 % compared with AA7075, respectively. The minimum micro hardness at SZ for AA7075+SiC surface nanocomposite records the lowest SD and SE of 1.18 HV and 0.68 HV, respectively. This implies the reinforcement is properly distributed within the surface of AA7075. Overall, carbonaceous reinforcements play a vital role in improving micro hardness than ceramic reinforcements.

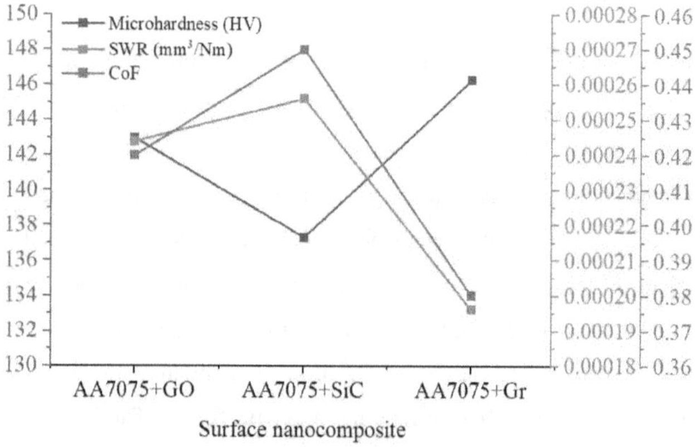

Fig.6.12 Micro hardness versus SWR and COF of AA7075 surface nanocomposites tested at 150^{0}C

The Gr reinforcement improves micro hardness by 6.53 % and 2.30 %, respectively, in comparison with SiC and GO. The SWR and COF at 150^{0}C tested condition are compared with the micro hardness of fabricated surface nanocomposites, which are depicted in Fig.6.12. The hardness of the AA7075 surface nanocomposite is in inverse relation to SWR. Better wear resistance is observed when the SWR is low. The minimum SWR yields low COF. For all fabricated surface nanocomposites, the SWR and COF are lower due to the enhancement in surface hardness. Among fabricated surface nanocomposites, AA7075+Gr surface results in a minimum SWR of 0.000196 ± 0.0000221 mm^3/Nm.

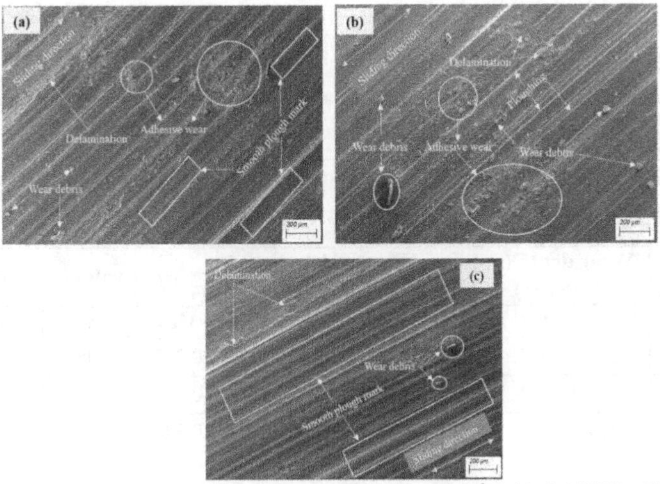

Fig.6.13 SEM morphology of worn surfaces at 150^{0}C (a) AA7075 +GO (b) AA7075+SiC (c) AA7075+Gr

Compared with SiC and GO reinforcement addition, Gr reinforcement improves SWR by 30.61 % and 24.48 %, respectively. Experimental results show that carbonaceous reinforcement yields better wear resistance than ceramic reinforcement. The improvement in wear resistance is attributed to the self-lubricating property of carbonaceous reinforcement particles (Liu et al. 2022).

The SEM wear morphologies are shown in Fig.6.13 (a-c). AA7075+GO worn surface shows the minimum size of wear debris, and it is depicted in Fig.6.13 (a). The size of wear debris is large in the AA7075+SiC worn surface, shown in Fig.6.13 (b). The cluster of wear debris having variable sizes denotes the dominating wear mechanism as an adhesive (Jamali and Mirsalehi 2022b). The predominant wear mechanisms are adhesive and delamination for AA7075+GO and AA7075+SiC surface nanocomposites. The surface of AA7075+Gr is smooth compared with other worn surfaces, as shown in Fig.6.13 (c). Most surfaces have smooth ploughing marks in the sliding direction and experience delamination. So, the wear mechanism of AA7075+Gr worn surface is a combination of abrasive and delamination (Murugabalaji et al. 2024a). The details of wear track of worn out surface nanocomposites are displayed in Table 6.12. Among reinforcement combinations, carbonaceous reinforced AA7075 surface nanocomposite exhibited minimum average surface roughness, wear depth, wear width and wear volume, unlike ceramic reinforcement. In particular, AA7075+Gr surface nanocomposite showed the lowest R_a, wear depth, wear width and wear volume of

0.018 ± 0.001 mm, −0.0042 ± 0.0012 mm, 0.1644 ± 0.0648 mm and 0.235 ± 0.018 mm^3, respectively. This performance suggests better resistance to material removal, most likely due to the Gr reinforcement's load-bearing ability and efficient solid lubricating effect.

Table 6.12 Details of the wear track of worn AA7075 surface nanocomposites tested at 150°C

Combination	Wear depth (mm) ($\times10^{-3}$)			Wear width (mm) ($\times10^{-1}$)			Wear volume (mm^3)		
	Mean	SD	SE	Mean	SD	SE	Mean	SD	SE
AA7075+GO	−5.0	3.8	2.2	3.434	0.912	0.527	0.2673	0.0582	0.0336
AA7075+Gr	−4.2	1.2	0.7	1.644	0.648	0.374	0.2346	0.0180	0.0104
AA7075+SiC	−5.6	3.4	2.0	3.611	0.514	0.297	0.3344	0.0786	0.0454

Figure 6.14 (a, b) shows the 3D surface roughness profile from WLI, wear depth and wear width plot of the worn-out AA7075+Gr surface. In contrast, AA7075+SiC surface nanocomposite shows the highest average surface roughness (0.023 ± 0.004 mm), wear depth (−0.0056 ± 0.003 mm), wear width (0.3611 ± 0.0514 mm) and wear volume (0.334 ± 0.079 mm^3). This result indicates a 27.78 % hike in R_a relative to the Gr reinforcement. Also, there is a notable hike in wear depth, wear width and wear volume. higher wear of the AA7075+SiC surface nanocomposite may be caused by interfacial debonding or micro-fracture between the reinforcement and matrix material (Kumar and Kumar 2024). Nonetheless, in some cases, increasing the SiC addition exhibited a reduction in both wear width and volume (Zhou et al. 2021). Fig.6.15 (a,b) shows the 3D surface roughness profile from WLI, wear depth and wear width plot of the worn-out AA7075+SiC surface. AA7075+GO composite displayed intermediate wear behaviour (R_a = 0.020 ± 0.004 mm, wear depth = −0.0050 ± 0.0038 mm, wear width=0.3434 ± 0.0912 mm, wear volume = 0.2673 ± 0.0582 mm^3) owing to its structural strength, layered structure and solid lubricant. Fig.6.16 (a, b) shows the 3D surface roughness profile from WLI, wear depth and wear width plot of the worn-out AA7075+GO surface. As per the wear performance trend (Gr < GO < SiC), a 2D carbonaceous nano reinforcement showed better in terms of reducing friction and protecting the surface when high-temperature dry sliding conditions than ceramic reinforcement.

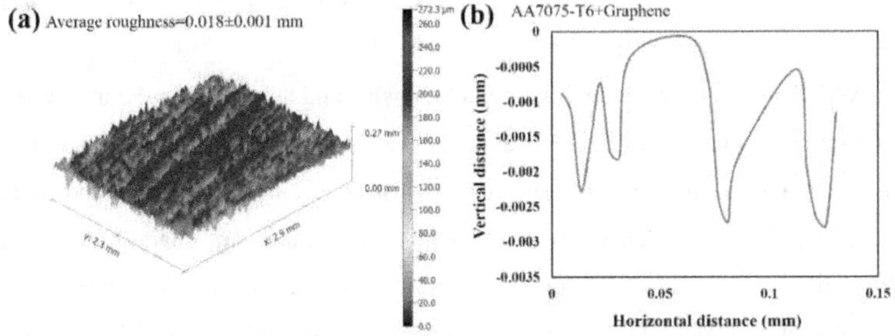

Fig.6.14 Surface topography of worn AA7075+Gr surface at 150^{0}C (a) 3D surface roughness profile (b) Wear depth and width

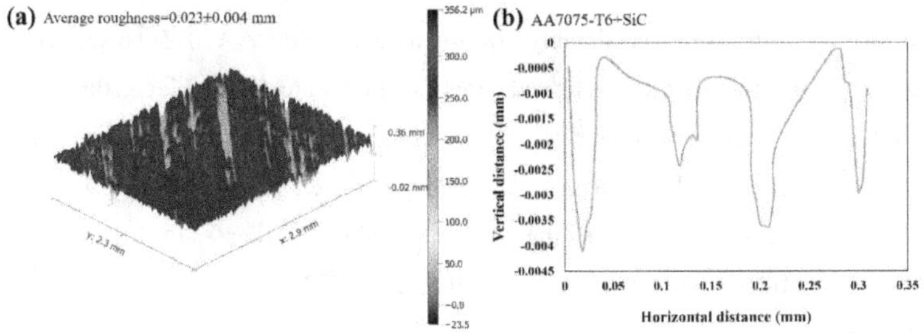

Fig.6.15 Surface topography of worn AA7075+SiC surface at 150^{0}C (a) 3D surface roughness profile (b) Wear depth and width

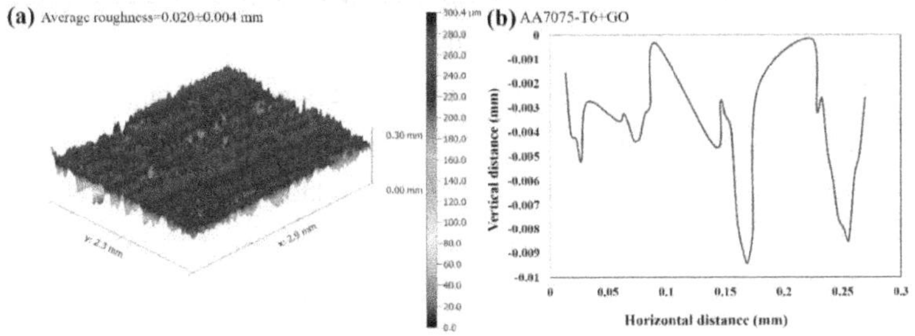

Fig.6.16 Surface topography of worn AA7075+GO surface at 150^{0}C (a) 3D surface roughness profile (b) Wear depth and width

6.4 COMPARISON OF TRIBOLOGICAL PERFORMANCE OF AA7075 SURFACE NANOCOMPOSITES

AA7075-T6 surface undergone severe abrasive and adhesive wear during room temperature dry sliding test (Deore et al. 2019). SEM images showed deep grooves, delamination, and material transfer caused by strong adhesion with the counter surface (Kumar et al. 2022; Patle et al. 2019). The soft aluminium matrix and coarse precipitates ($MgZn_2$) promoted formation of crack, leading to high friction, unstable sliding, and significant material loss (Patil et al. 2022). In contrast, FSPed AA7075 (without reinforcement) produced shallower grooves and smoother wear tracks. The grain refinement caused by FSP enhances hardness and load-bearing capacity, and reduces plastic deformation (Nathan et al. 2023b). As an outcome, adhesive wear decreased and the wear mechanism shifts to a mild adhesive and abrasive type with stable sliding. The wear rate and COF were considerably lower than those of the AA7075-T6 surface, mainly due to refined microstructure, closed pores, and better bonding achieved through FSP. AA7075 alloy showed adhesive wear mechanism at elevated temperature condition (Murugabalaji et al. 2024a). Similarly, unreinforced FSPed AA7075 at 140^{0}C showed adhesion wear mechanism as the material of pin was transfer to the counter face, owing to an effect of temperature (Loganathan et al. 2021).

The tribological behavior of AA7075-T6 with unique combinations of reinforcements (SiC, GO, GNPs, and Gr) under different processing conditions are displayed in Table 6.13. The results exhibited that the wear depth and volume were enhanced significantly at elevated temperature dry sliding tests (150°C) compared to the room temperature dry sliding tests. Under room temperature dry sliding test, the wear depth lied between −0.0013 mm and −0.0037 mm, and wear volume differs from 0.221 mm^3 to 0.371 mm^3. In contrast, at elevated temperature dry sliding test, wear depth fluctuates between −0.0042 mm and −0.0056 mm, with wear volume varies from 0.235 mm^3 to 0.361 mm^3. This enhancement in wear at higher temperatures can be attributed to the softening of the material, which makes it more susceptible to wear. The COF was found to be enhanced at high temperature, with values fluctuating from 0.38 to 0.45, compared to 0.23 to 0.29 at room temperature condition. This shows that, as the material softens at elevated temperature, the friction between the sliding surfaces enhances, leading to larger SWR. SEM analysis of the worn surfaces confirmed that, under room temperature, the wear mechanism was primarily abrasive, with consistent grooves and minimal debris, showing that the material's hardness contributed to resist significant wear. At elevated temperature, adhesive, delamination, and ploughing were more prominent wear mechanisms, with deeper grooves and larger debris formation, suggesting that the material was more prone to deformation. In terms of nano reinforcement, ceramic particles were found to significantly improve wear resistance, notably under room temperature, where the AA7075+SiC combination produced the best performance. The findings are consistent with previous work, that SiC nano reinforcement improved the wear resistance of AA7075 composites (Suresh et al. 2019). Gr reinforcement provided lubrication effect and reduced COF, especially under room temperature conditions. This align with the work of Namdev Ashok Patil et al., who found that graphite particles reduced friction in aluminium alloys (Patil et al. 2022).

The enhance in wear at high temperatures is in line with the outcomes of Murugabalaji et al., who observed that wear rates enhanced in AA7075 composites when tested at elevated temperatures (Murugabalaji et al. 2024a). This enhance in wear is attributed to the softening of the material, which makes it less effective at resisting sliding contact, promoting to more wear and a higher COF. The existing work also confirmed this trend, as the AA7075-T6 material softened and lost its resistance to wear at higher temperatures. This research also highlights the significance of reinforcement selection and processing variables for optimizing the wear resistance of AA7075-T6. While SiC reinforcement improved wear resistance under room temperature, it is less effective at elevated temperatures, where the material undergoes significant softening. Gr, although effective at reducing COF under room temperature, may not be as effective at high temperatures. The comparison of present findings with existing works especially elevated temperature dry sliding wear test are summarized in Table 6.14. With reference to existing works, the tribological performance of fabricated AA7075 surface nanocomposites were found be the enhanced.

Table 6.13 Comparison of room and elevated temperature dry sliding wear performance of AA

Case	Combination	CS (nm)	Micro hardness	SWR (mm^3/Nm)	COF	Wear depth (mm)	Wear width (mm)	Wear volume (mm^3)	Ra (μm)
1	SiC+20	56.723	127	0.000194	0.24	−0.0024	0.205	0.329	0.024
	GO+25	47.196	130	0.000127	0.23	−0.0013	0.097	0.221	0.017
	GNPs+30	66.463	117	0.000232	0.29	−0.0037	0.527	0.371	0.029
2	AA7075+GO	52.972	142.97	0.000241	0.42	−0.0050	0.343	0.267	0.021
	AA7075+SiC	56.235	137.3	0.000261	0.45	−0.0056	0.361	0.334	0.023
	AA7075+Gr	45.212	146.27	0.000196	0.38	−0.0042	0.164	0.235	0.018

Table 6.14 Comparative analysis of the present research work with existing works under elevated

Ref	AA7075 reinforced with	Processing method	Wear test conditions				Comparis
			P (N)	SD (m)	V (m/s)	T (^{0}C)	
(Robinson Smart et al. 2020)	$TaC+Si_3N_4+Ti$	Stir casting	20-40	3000	2-4	30,60 and 120	This study shows lo conditions (120°C, 20 temperature range com and 2 m/s), where grap
(Shivapraka sh et al. 2023)	2wt.% granite powder+Si_3N_4	Stir-casting and Heat treatment	20-50	1000	0.419	35-210	The wear rate and temperature. At a no composites was consis mode changed to ab current findings.
(Shin et al. 2019)	B_4C	Liquid pressing	40	200	-	RT, 100, 200, 300	The inclusion of enhancement in wear 300^{0}C) owing to the fo Similarly the present resistance at 150^{0}C fo wear width and depth Interestingly, wear delamination) formati work was found same

P-applied load, SD-Sliding distance, V-Sliding velocity and T-Temperature

Table 6.14 Comparative analysis of the present research work with existing works under elevated

(continues..)

Ref	AA7075 reinforced with	Processing method	Wear test conditions				Compa
			P (N)	SD (m)	V (m/s)	T (^{0}C)	
(Singh et al. 2019)	Al_2O_3+Graphite	Stir casting	15,20, 25	1000	-	30, 115, 200	This work exhibited affected by temperatu by the formation of lower wear rate at 15(graphite combination
(Murugabalaji et al. 2024a)	SiC+Graphite	Stir casting, Rolling and annealing	30,50, 70	-	1.062, 1.366, 1.669	30, 250, 300	The hot cross rolled hy owing to the load be lubricating ability o graphite in both works study operates at a lo different wear mechar
(Michael Rajan et al. 2014)	TiB_2	In-situ reaction	20	1000	1.2	30-240	The inclusion of TiB_2 wear rate increase wit reported at 150^{0}C s suggesting that graph lubricate more than Ti

P-applied load, SD-Sliding distance, V-Sliding velocity and T-Temperature

6.5 SUMMARY

The chapter investigated the wear behaviour of AA7075 surface nanocomposites through pin-on-disc tribometer under varying conditions. Statistical analysis including descriptive statistics, normality, non-parametric tests, and ANOVA were applied to interpret the specific wear rate data. The worn samples were further examined through microhardness, surface morphology and topography analyses to understand wear mechanisms. The summary points are as follows.

- ANOVA confirmed that both tool traverse speed and reinforcement were strongly influence dry sliding wear behavior under different conditions.
- The GO+25 combination exhibited the lowest specific wear rate, highest hardness, and lowest coefficient of friction. A clear inverse correlation between micro hardness and specific wear rate was established, confirming strengthening induced wear resistance.
- SiC reinforcement predominantly exhibited abrasive wear, whereas GO reinforcement favoured adhesive wear with minimal debris, indicating smoother sliding behavior. Surface topography confirmed that GO+25 resulted in minimum wear depth, width, volume, and surface roughness, reflecting superior integrity under load.
- At elevated temperatures, reinforcement influence became statistically insignificant, suggesting thermal effects dominate wear mechanisms at high temperature. AA7075+Graphene demonstrated lowest specific wear rate and exceptional consistency at elevated temperatures, indicating better thermal wear stability.

CHAPTER 7

CONCLUSION

This research is focused on investigations of the surface integrity, corrosion and tribological behaviour of friction stir processed AA7075 surface nanocomposites. The following are research contributions and research future scope.

7.1 RESEARCH CONTRIBUTIONS

- A systematic literature review is performed on improving AA7075-T6 alloy surface properties using friction stir processing (FSP) with carbonaceous and ceramic reinforcements. The research on the friction stir processed (FSPed) aluminium alloy and its mechanical, tribological, and corrosion behaviour are reviewed and presented. The role of FSP processing variables on temperature distribution, microstructure, mechanical and surface properties are reviewed and presented.
- The fabrication of AA7075 surface nanocomposites were done using varying tool pin profiles and traverse speeds. The processing temperature during fabrication was monitored. Carbonaceous and ceramic nanoparticles were served as reinforcements. The FSPed samples were characterized using different techniques. The behaviour of surface integrity, corrosion and tribology of AA7075 surface nanocomposites were evaluated as per testing standards.
- SiC+25 combination exhibited best surface integrity behaviour, in terms of refined crystallite size, lowest surface roughness, enhanced ultimate tensile strength and micro hardness. A tool traverse speed of 25 mm/min facilitated, enhanced processing temperature and increased compressive residual stress, lead to better material flow and dynamic recrystallization.
- ANOVA outcome implies reinforcement has significantly impacted on corrosion rate, unlike tool traverse speed. An addition of SiC refined the microstructure of AA7075 surface. A refined crystallite size (47.162 nm) of SiC+25 surface, lead to enhance the proportion of grain boundaries (GBs), which declined the corrosion resistance owing to the higher energy and chemical activity of GBs. A pitting corrosion is evident for all combinations, GO+30 revealed shallower pits with R_a of 136.5 nm, unlike other combinations.
- ANOVA outcome implies both tool traverse speed and reinforcement were equally contributed on determining specific wear rate (SWR). The wear depth

and volume were enhanced significantly in the elevated temperature dry sliding tests (150°C) compared to the room temperature dry sliding tests. Both room and elevated temperature dry sliding wear condition, surface topography analysis revealed that carbonaceous reinforced AA7075 surface produced least wear depth, wear width, wear volume and surface roughness.

7.2 SCOPE FOR FURTHER RESEARCH

- Study the influence of dual reinforcements to enhance surface integrity and stress corrosion cracking
- Analyze the synergies between carbonaceous and ceramic reinforcements for multi-functional properties such as electrical and thermal conductivity.
- Study the influence of advanced tool pin geometries, rotational speeds, and traverse speeds on the dispersion and bonding of reinforcements
- Explore the potential of combining FSP with advanced surface coating techniques to further improve surface hardness, wear resistance, and corrosion resistance of the Aluminium surface nanocomposites.

www.ingramcontent.com/pod-product-compliance
Lightning Source LLC
LaVergne TN
LVHW010726061126
851857LV00014B/1369